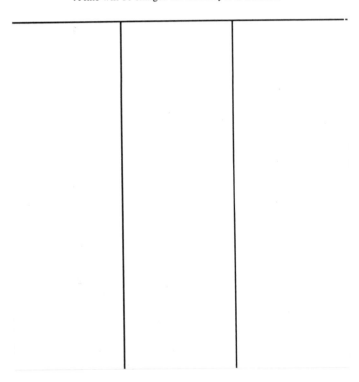

D1357211

Pest and Vector Control

As ravagers of crops and carriers of diseases affecting plants, humans and animals, insects present a challenge to a growing human population. In *Pest and Vector Control*, Professors van Emden and Service describe the available options for meeting this challenge, discussing their relative advantages, disadvantages and future potential. Methods such as chemical and biological control, environmental and cultural control, host tolerance and resistance are discussed, integrating (often for the first time) information and experience from the agricultural and medical/veterinary fields. Chemical control is seen as a major component of insect control, both now and in the future, but this is balanced with an extensive account of associated problems, especially the development of pesticide-tolerant populations.

Pest and Vector Control

H. F. van Emden

Emeritus Professor of Horticulture
School of Plant Sciences
The University of Reading

and

M. W. Service

Emeritus Professor of Medical Entomology
Vector Research Group
Liverpool School of Tropical Medicine

CAMBRIDGE
UNIVERSITY PRESS

PUBLISHED BY THE PRESS SYNDICATE OF THE UNIVERSITY OF CAMBRIDGE
The Pitt Building, Trumpington Street, Cambridge, United Kingdom

CAMBRIDGE UNIVERSITY PRESS
The Edinburgh Building, Cambridge CB2 2RU, UK
40 West 20th Street, New York, NY 10011-4211, USA
477 Williamstown Road, Port Melbourne, VIC 3207, Australia
Ruiz de Alarcón 13, 28014 Madrid, Spain
Dock House, The Waterfront, Cape Town 8001, South Africa

http://www.cambridge.org

© H. F. van Emden and M. W. Service 2004

First published 2004

Printed in the United Kingdom at the University Press, Cambridge

Typefaces Adobe Garamond 11/13 pt. and Frutiger   *System* LATEX 2$_\varepsilon$ [TB]

A catalogue record for this book is available from the British Library

Library of Congress Cataloguing in Publication data

Van Emden, Helmut Fritz.
Pest and vector control / H.F. van Emden and M.W. Service.
  p. cm.
Includes bibliographical references and index.
ISBN 0 521 81195 3 – ISBN 0 521 01083 7 (pb.)
1. Pests – Control. 2. Pests – Biological control. 3. Vector control.
4. Pesticides – Environmental aspects. I. Service, M.W. II. Title.
SB950.V35   2003
632′.7–dc21   2003053198

ISBN 0 521 81195 3 hardback
ISBN 0 521 01083 7 paperback

Contents

Preface

For nearly 30 years, generations of students of crop protection have used a slim volume, written by one of us (HvE), and first published in the *Studies in Biology* Series entitled *Pest Control and its Ecology* (1974) and later revised with the title simplified to *Pest Control* (1989).

When the time came for a 3rd edition, the publisher (in the form of Ward Cooper of Cambridge University Press) asked that the book be enlarged and expanded to include areas of applied entomology not included previously, particularly the control of insects of medical and veterinary importance.

Fortunately we had been undergraduates together in the Department of Zoology and Applied Entomology at Imperial College, graduating in 1955 and, although agricultural and medical entomology led our respective careers in different directions immediately thereafter, we have remained in contact and firm friends ever since. The co-authorship of the new enlarged edition was therefore never in doubt!

Like Pest Control (1989), this book is also limited to the control of arthropods; we felt that amplifying the title would make the latter cumbersome if more descriptive.

We think the result is a book unique in the width of its coverage of the control of problem insects. We have not only covered insects of agricultural, stored product, medical and veterinary importance, but we have included the full range of control methods, including some which will be unfamiliar to most readers. These follow a general introduction on how insects interact with man and a 'rough guide' to the essentials of animal population dynamics as necessary to understand how insect problems arise. In then going through the different control methods, we give our opinion on their advantages and limitations as well as their relative importance and where they are going in the future. Even those still on the research bench, and those we suspect may not be going anywhere, are included. This is because we wish to explore the rich variety of man's ingenuity in his battle against insects and make it clear that contributions have come from unexpected quarters, e.g. the physics of spectral absorption of different glasshouse cladding materials and the design of machines for paint-spraying metal grids. Another rather unusual feature of our book is that we not only include the components and principles of pest management but, in our final chapter, also explain how the different components may be combined and integrated into pest management programmes.

Now anyone in either the agricultural or medical entomology field will know that 'never (or only very rarely) the twain shall meet'. Conferences or day-meetings on the two entomological disciplines attract totally different audiences, who hardly read each other's textbooks or scientific papers. We are ourselves examples of this; we believe we have never attended the same meeting or conference. Even the indispensable applied entomology abstracting journal, the *Review of Applied Entomology*, was split into two distinct annual volumes (*Agricultural* and *Medical and Veterinary*) as long ago as 1913.

Combining the two areas of entomology in a single book has previously rarely been attempted, and we quickly discovered a major difference as to how pest control is subdivided in our two disciplines. In agriculture there are many crops, with several major pests on each; control is usually practised on the clearly defined area of the crop. In medical/veterinary entomology, by contrast, the types of problems are fewer, but nevertheless some of the problems involve really serious diseases transmitted by arthropods to very large populations of humans or livestock. Control of the vectors often is not on the attacked target (a human or animal) but carried out in the wider environment of that target, an environment which is usually heterogeneous and may be on a very large (e.g. regional or countrywide) scale.

The result of these contrasts is that, whereas a text on pest control in agricultural entomology is divided by control approach (chemical, biological, cultural etc.), control in medical/veterinary entomology is usually focused on the several different methods needed for control of a particular disease (e.g. control of malaria, sleeping sickness), and then how much each control method contributes.

We took the decision to follow the agricultural model and integrate into this approach the relevant examples from the medical/veterinary field. Nevertheless, some topics proved impossible to treat in this way. So there are, for example, separate sections in Chapter 13 for the two disciplines on thresholds and insect monitoring and forecasting. The reader will quickly find other such examples.

However, one advantage of trying to integrate our material is that the links of agriculture with human and veterinary diseases are easy to recognize, and we have stressed these links wherever possible. For example, increased rice cultivation to feed an increasing population inevitably creates places for mosquitoes to breed, and intensive and extensive spraying of cotton with insecticide can sometimes result in insecticide resistance in malaria vectors.

We have had some difficulty in knowing how we should deal with the various active ingredients of insecticides. These chemicals are currently under intense scrutiny in relation to safety to human health and the environment; many have been banned or withdrawn by the manufacturer. Unfortunately, the position changes almost daily and differs between countries. Anti-cholinesterase compounds (particularly the organophosphates) are primary targets for this scrutiny, yet of all the chemical groupings they are the best example of the variety of routes to the target. We have therefore mentioned the compounds which best illustrate the properties of insecticides and the variations found between active ingredients. We hope we have not suggested that any compound universally banned is still available, while the corollary is that mention of a chemical in this book cannot be taken to mean that it is available and recommended for use for all situations, and in whatever country the reader is located.

Repetitions, and exactly where subject matter is treated and in which chapter, are always problematic with a book of this kind. For example, genetically engineered crops expressing the *Bacillus thuringiensis* toxin are an example of genetic manipulation (Chapter 9) which also represents a delivery system for an insecticide (Chapter 4) which is derived from an insect pathogen (Chapter 8) and gives the crop plant resistance to insect attack (Chapter 11) with implications for pest management (Chapter 13)! Where necessary we have accepted some repetition, but have indicated where a more extensive treatment of the topic can be found in the book. In other places, we have attempted to explain why a topic is not discussed there, again pointing out the relevant chapter.

Chapter 12 needs some comment. As well as bringing together a miscellany of insect control methods for which there was no obvious home elsewhere, we have a section on 'Other topics'. These are not methods of control, but are relevant to such methods. There are legal requirements to control some insects or prevent their spread – topics we do regard as insect control – and also legislation on, for example, which insecticides may be used and how they may be used – this we do not regard as insect control, but it is clearly highly relevant. Similarly, controlling mosquitoes is clearly insect control, yet it is only a part of the management of malaria and so we felt it not inappropriate to mention briefly the use of drugs, a very important component of malaria, but not of mosquito, control. Involving the local community in what control is required and how best to implement it (community participation) is a further 'Other topic'.

The earlier editions of this book referred to at the start of this Preface gave guidance to general reading. In this volume, additional literature sources have been mentioned in the text, sometimes because we have taken a table, figure or quotation from that source. For the sake of simplicity, we have combined the literature cited in the text into one bibliography which also contains our suggested general reading, usually books or reviews, not mentioned in the various chapters.

We have enjoyed working together on the book, and have benefited greatly from learning much more about each other's discipline. We have relied greatly on our own experience during our careers and information acquired during discussions with colleagues and at conferences. Our aim has been to keep the book readable, hoping our enthusiasm for the subject permeates its pages. We have therefore not held back from including stories we enjoy or find bizarre, even if these make the balance of detail given to different topics somewhat unequal.

1

Man and insects

Impact on man

Certain insects, such as flies, lice and locusts have been associated with ill health of humans, sickness in domesticated animals and crop losses from pre-biblical times. However, since man was a hunter-gatherer long before he took up farming, his first experience of insect problems must have been with being bitten by them. A much-quoted biblical passage describes the plague of flies the Lord is said to have despatched to Egypt which entered the house of Pharaoh (Exodus). In spite of this long association, connections between biting insects and disease took many centuries to be made, whereas the depredations of insects on crops were largely instantly recognizable. The eccentric physician Erasmus Darwin, grandfather of Charles, came very close to guessing the truth that lice transmitted typhus, but it was only during the late nineteenth and early twentieth centuries that insects such as mosquitoes were identified as vectors of malaria, yellow fever and other infections, that tsetse flies transmitted sleeping sickness and animal trypanosomiasis, and that ticks spread various infections, such as so-called Texas fever, to cattle (Table 1.1). Similarly the first proof that an insect (the honey bee) transmitted a disease of plants (fireblight) was not obtained until 1892. Although the pace of vector incrimination has slowed down, more recent discoveries have identified ticks as vectors of the spirochaetal organisms causing Lyme disease in humans.

Mosquitoes have had a drastic effect on man's progress. For example, anopheline mosquitoes are responsible for transmitting the best known and most important, and arguably man's oldest, vector-borne disease, namely malaria. It is probable that the disease originated in Africa, and followed

Table 1.1. *Some major infections transmitted by arthropods.*

Vector	Parasite	Disease	Main distribution*
Aphids			
Bird-cherry aphid	Virus	Barley yellow dwarf	Worldwide
Peach-potato aphid	Virus	Beet yellows	Worldwide
	Virus	Potato leaf roll	Worldwide
Brown citrus aphid	Virus	Citrus tristeza	Subtropics
Leafhoppers			
Green rice leafhopper	Virus	Rice tungro	Asia
Maize leafhopper	Virus	Maize streak	Africa
Beet leafhopper	Virus	Beet curly top	Western USA, Mexico, Canada, Mediterranean
Planthoppers			
Brown planthopper	Virus	Rice grassy stunt	Africa, Asia
Maize planthopper	Virus	Rice stripe	Asia
Plant bugs			
Cocoa capsid	Virus	Swollen shoot (cocoa)	Africa
Antestia bug	Fungus	*Nematospora* taint of coffee	Africa
Bees	Bacterium	Fireblight	North and South America, Europe, New Zealand
Bark beetles	Fungus	Dutch elm disease	Asia, Europe, USA, Canada
Mosquitoes			
Anophelines	Protozoa	Malaria	Tropics, subtropics
	Filaria	Filariasis	Tropics
Culicines	Arboviruses	Yellow fever	Africa, South America
		Dengue	Tropics
	Numerous other arboviruses		Worldwide
Tsetse flies	Protozoa	Human and animal trypanosomiasis	Africa

Table 1.1. (*cont.*)

Vector	Parasite	Disease	Main distribution*
Simuliid black flies	Filaria	Onchocerciasis	Africa, South America
Sand flies	Protozoa	Leishmaniasis	Tropics, subtropics
Muscid flies	Protozoa	Enteric infections	Worldwide
	Bacteria etc.	Mastitis	Worldwide
Triatomine bugs	Protozoa	Chagas disease	Central and South America
Fleas	Bacteria	Plague	Worldwide
Body lice	Rickettsiae	Epidemic typhus	Worldwide
	Spirochaetes	Epidemic relapsing fever	Tropics, subtropics
Argasid ticks	Spirochaetes	Tick-borne relapsing fever	Tropics, subtropics
	Rickettsiae	Aegyptianellosis	Worldwide
Ixodid ticks	Arboviruses	Tick-borne encephalitis	Europe
		Colorado tick fever	North America
		Many other arboviruses	Worldwide
	Rickettsiae	Many typhuses	Tropics, subtropics
	Spirochaetes	Lyme disease	Worldwide
	Protozoa	Theileriosis	Worldwide
		Babesiosis	Worldwide

* The distributions are only approximate, for example worldwide indicates the infection spans the tropical, subtropical and temperate regions, but is not necessarily widespread in all or any region.

in the wake of human migrations out of Africa to the Mediterranean, the Indian subcontinent and South-East Asia. Malaria is said to have caused, or aided, the decline of the Roman Empire and fall of Greece; how true this is remains debatable. Nevertheless, malaria has certainly helped defeat armies involved in military campaigns, sometimes causing more deaths than military action. For example, during the First World War (1914–18) malaria outbreaks immobilized armies in Macedonia for three years. During the Vietnam war (1965–73) the American army had as many soldiers suffering from malaria as battle casualties. More recently, in 1988, the anopheline mosquito caused more than 25 000 deaths in Madagascar. Malaria has played a major role in retarding the cartographic exploration of Africa, and inhibiting trade between Africa and Europe. Malaria was mainly responsible for West Africa being known as 'The White Man's Grave', because it claimed so many lives at the beginning of the nineteenth century. Statistics can be notoriously inaccurate, but an educated guess is that malaria currently kills about 2 million people annually, some 90% of this occurring in sub-Saharan Africa.

Because of the so-called yellow fever mosquito, *Aedes aegypti,* construction work on the Panama Canal was abandoned in 1889 due to the devastation this mosquito caused in transmitting yellow fever to the itinerant work force. Some 20 000 workers, comprising young engineers, administrators and labourers were killed. And in 1895–1908 there were at least 28 000 deaths from yellow fever before efficient control measures against *Aedes aegypti* were adopted. These days there is an excellent vaccine that provides 10, or more, years of protection.

In some parts of the world, such as subarctic regions of North America, mosquitoes can be so numerous as to prevent just about all outdoor activities. In fact research on mosquito biology and control has been financed by the Canadian government because their armed forces would find it difficult to defend certain terrain because of the intolerable nuisance caused by mosquitoes.

Tsetse flies, *Glossina* species, are found only in Africa, but they have greatly influenced the development of that continent. They transmit two related protozoal diseases, sleeping sickness (human trypanosomiasis) and animal trypanosomiasis (often called nagana).

The animal disease infects domestic livestock, especially cattle, which become emaciated, sick and often die. Areas where tsetse flies are particularly numerous are often called fly-belts, and cattle owners try to avoid these areas. The disease is found in 37 sub-Saharan countries covering some 11 million km^2 of land, and causes an estimated annual loss of US\$ 5000 million. Animal trypanosomiasis has greatly hindered African agricultural development and kept communities poor.

Insects have attacked crops since the dawn of agriculture. Chinese cave paintings dating about 4000 BC and Egyptian artefacts from 2300 BC depict pests attacking crops. The ancient Egyptians were also only too aware that food brought into store was rapidly destroyed by insect and rodent pests. Close on the biblical plague of flies referred to earlier, the Lord visited a plague of locusts on the Egyptians. ' . . . and when it was morning, the east wind brought the locusts . . . they covered the face of the whole earth, so that the land was darkened; and they did eat every herb of the land, and all the fruit of the trees . . . ; and there remained not any green thing in the trees, or in the herbs of the field, through all the land of Egypt' (Exodus). Today swarms of locusts still darken the skies of Africa. These swarms may occupy hundreds of square kilometres of air space with 100 million individuals (weighing more than 100 tons) in every km^2. Swarms can travel hundreds of kilometres a day and, as stated in Exodus, they destroy every piece of green vegetation where the swarm lands.

Aphids are another major world problem. They neither swarm nor chew, but drift as isolated individuals in the air and then suck the sap of the plants they attack. Numbers can again be huge (e.g. over 200 million black bean aphids (*Aphis fabae*) per hectare in sugar beet), arrived at by multiplication from perhaps only 500 colonizing individuals.

The great Chinese famine (1958–62), in which 30 million people starved to death, was in large part due to uncontrolled pest populations, following an amazing contrived reduction in their bird predators. In an attempt to control sparrows, huge numbers of Chinese were mobilized to make noises by beating metal pans etc. to prevent sparrows landing till the birds died of exhaustion.

In the spring of 2001 there was a plague of armyworms which again devastated crops in the eastern and southern provinces of Cameroon, threatening a famine like the one that followed a similar event in 1980. One plantation owner was quoted as follows: 'Not a single green plant is spared, from bananas to cocoyams to groundnuts.'

The first succinct treatise on agricultural entomology is again in the Bible, in the book of the prophet Joel. 'That which the palmerworm hath left hath the locust eaten; and that which the locust hath left hath the cankerworm eaten; and that which the cankerworm hath left hath the caterpillar eaten.' Many attempts have been made to measure the overall losses of our crops to insects. These estimates differ considerably. One estimate which several workers have reached is that we grow about a third of our crops to feed insects either in the field or in store, in spite of modern control measures being taken to minimize such losses where we can afford to do so. Put another way, we grow plants as insect food on something approaching the equivalent of 50 million hectares!

Today there are crop pests which take their place alongside mosquitoes and tsetse flies on the world stage as ringleaders on the insect side of what is sometimes called the 'Insect War'. Such are locusts, grasshoppers, armyworms, bollworms of cotton, diamond-back moths (*Plutella xylostella*) of brassicas and brown planthoppers (*Nilaparvata lugens*) of rice. More detail of these major pests will be given elsewhere in this chapter, but the last three are typical of the many pests which have dramatically changed (in this case increased) in importance since the advent of insecticides, the introduction of new crop varieties and more intensive agriculture. They will appear as the targets of pest control measures in other chapters, but would probably hardly have warranted mention had this book been written in the 1950s. Man has created many new pests by his efforts to control others and to strive for ever higher crop yields. Thus insecticides used against one pest of apples destroyed the biological control of others (perhaps most famously the fruit tree red spider mite, *Panonychus ulmi,* in the UK). Similarly, trying to control the one and only pest scale insect of citrus (an imported species) in California at the start of the twentieth century with insecticide has raised the number of scale insect species which now may need control to nearly 100.

Colonization of Africa by Europeans is believed to have caused the spread of sleeping sickness. There have been disastrous epidemics and deaths, due to the opening up of trade routes that allowed African populations to travel into what were formerly hostile areas. David Livingstone and his large entourage of porters, some already infected, are credited with spreading sleeping sickness in parts of eastern Africa. There were dramatic epidemics in the early 1900s. Although sleeping sickness still kills at least 20 000 people a year, the greatest economic burden for Africa is the animal form of trypanosomiasis.

Human travelling and trade have moved crop pests and crop plants around the world and, like the movement of insect-vectored human diseases, many new problems have been caused thereby. Some examples are given later under 'migrant pests'. The colonization of North America by Europeans created a major new pest problem, not because an insect but a new crop was introduced to the USA. In 1824, a British insect collector visited the Rocky Mountains, and returned with an attractively striped beetle which was new to science and which he named *Leptinotarsa decemlineata.* He had found sparse populations of the beetle feeding on a weed (buffalo-bur) on the eastern slopes of the mountain range. Thirty years later, settlers from Europe started growing potatoes (the same plant family as buffalo-bur) on the plains and, when they reached the foothills of the Rockies, the beetle transferred to this new nutritious and almost limitless food resource. The new pest spread eastwards towards the Atlantic seaboard at 140 km a year, and soon travelled in the opposite direction to the human settlers to reach Europe. Today the beetle, under the name 'Colorado beetle' (Fig. 2.6) is a serious pest of potatoes in much of Europe and figures in warning posters on police station notice boards in the company of wanted criminals.

Fleas and lice from time immemorial have changed the demography of the world. Fleas, mainly the tropical rat flea, *Xenopsylla cheopis,* have been responsible for three pandemics of bubonic plague, namely the ancient plague which culminated in the great plague of Justinian's reign (*c.* AD 542), the Black Death of the Middle Ages (1347–55) which killed 30–60% of the population of Europe, and an epidemic which is believed to have originated in China in about 1870, and which in 1895 on reaching India killed an estimated 1 300 000 people. Plague is reputed to have retarded western civilization by about 200 years. In addition to these pandemics there was the Great Plague of 1665–6 which killed more than 68 000 people in the UK, and the plague of Marseilles (1720–1) which killed nearly 40 000 of the city's 90 000 population.

Another ectoparasitic insect that has caused an immense loss of life is the body louse (*Pediculus humanus*). This louse transmits epidemic typhus, also formerly known as jail fever or ship fever because the overcrowded and unsanitary conditions that occurred in jails, ships and in army camps encouraged the spread of body lice. When Thomas à Becket was murdered in Canterbury Cathedral it is said that body lice marched from his cold body like a retreating army. The body louse caused appalling typhus epidemics in the sixteenth and seventeenth centuries. The potato famine in Ireland plus typhus outbreaks resulted in the migration of 1.5–2 million people to the USA from 1845 onwards. As recently as 1917 and 1923 some 30 million people in eastern Europe were infected with typhus; about 3 million of them died. At other times insects have caused nuisances by their sheer numbers. It has been said that the historically poor state of croft farming in western Scotland is due to multitudes of biting midges (*Culicoides* species) hindering farming activities in the evenings. During the eighteenth century the simuliid black fly (*Simulium colombaschense*), known as the Golubatz fly, was such a biting pest along the shores of the River Danube that cattle were killed in their hundreds and even thousands by sheer loss of blood! The last major outbreak of these black flies was in 1951 and killed 801 animals.

In the eighteenth and nineteenth centuries ticks were responsible for many cattle deaths in the USA due to their transmission of a protozoal parasite (species of *Babesia*) causing a disease known as Texas fever or redwater fever.

Famines have been a feature throughout history, and it is likely that crop pests have contributed equally to human misery as have the insects carrying diseases. Yet, although some 350 cases of famine can be found in historical records since Roman times, the records do not distinguish between pests, crop diseases or events such as drought as the cause. However, there are some relatively recent accounts of crop destruction. One of these concerns the total destruction of the first cereal crops established by the Mormons in 1847 when they settled in Utah. The insect was *Anabrus simplex*, now known as the 'Mormon cricket'. When the crop was threatened again the following year, the settlers gathered in church to pray for deliverance and, amazingly so far inland, a huge flock of seagulls appeared and ate up the crickets. This we believe to be the first and only example of biological control by divine intervention. In 2001, a plague of similarly devastating proportions occurred in Utah; this time prayers were replaced by the broadcasting of huge quantities of poisonous bait.

Pests and vectors

What is a pest?

Almost any animal can be a pest. For example, in Kenya many farmers regard elephants as pests because they often destroy crops. Honey bees are regarded as beneficial insects, but the African honey bee (*Apis mellifera scutellata*) can be a serious, and dangerous, pest because of its unprovoked attacks on people and livestock. It can be hard to give an all-inclusive definition of a pest. However, we might define an arthropod pest as any species that is injurious or potentially injurious to humans, their possessions, plants, plant products or domesticated animals. They may do this in a variety of ways, including disseminating diseases (see below).

Crop pests can damage any part of the plant above and below ground by chewing, sucking, tunnelling or causing plant tissue deformations known as 'galls'. Insects (e.g. termites, leaf-cutter ants) may remove parts of plants to build their nests and shelters and/or to take into the nest for food or to use a substrate for growing fungus gardens for later grazing. Insects can also be damaging by cross-fertilizing certain rust (fungal) diseases.

Pests may irritate or injure man and domestic animals by just being present, settling on the skin, by odours they may produce or by entering openings. They may sting, bite, suck blood or cause rashes and other allergies; also they may invade the body and tunnel into tissues (e.g. muscles) or enter organs including the alimentary canal.

What is a vector?

Some pests are such because, more importantly than any direct damage they may cause by their feeding, they are vectors, i.e. organisms that transmit infections. In the broad sense dogs can be vectors! By biting people they can transmit the rabies virus. A distinction is often made between biological vectors and mechanical vectors. Although we distinguish these two types of vectors, the disease rather than the insect determines the type of transmission. Thus the same aphid species may be the biological vector of one plant virus, but the mechanical vector of another. Any one disease is, with a few exceptions of medical and veterinary infections, transmitted in the same way. Among the exceptions is *Trypanosoma vivax,* which is cyclically transmitted by tsetse flies in Africa whereas in South America it is mechanically spread by tabanids and stable flies. The serious barley yellow dwarf virus disease is transmitted by at least ten different aphid species, but always in a biological manner. With biological vectors, the pathogens or parasites undergo a developmental cycle in the vector, or multiplication, or both. Examples are leafhoppers spreading Tungro virus disease of rice in Asia and mosquitoes transmitting malaria parasites. With mechanical transmission there is no multiplication or cyclical development in the vector, the infectious agents being passively carried by the vector. Examples include aphids spreading potato leaf roll virus and house flies transmitting enteric pathogens by their contaminated feet, vomit or faeces. Usually there are more direct methods by which such parasites are transmitted, such as unhygienic handling of food. With plant viral diseases, mechanical transmission can usually be reproduced with the point of a pin.

Some arthropods, especially ticks, can become infected with pathogens during a nymphal stage and transmit the pathogens when any subsequent nymphal stages and adults feed. This is termed transstadial transmission. Even more interesting and of greater epidemiological significance is transovarial transmission. This is when an infection in a tick, or a few other types of arthropods, passes to the ovaries which results in the nymphal and subsequent stages being infected and thus becoming vectors before they have even taken a blood-meal.

Most arthropods of veterinary or medical importance such as mosquitoes, lice, ticks and fleas are regarded as vectors, because of their role in the transmission of diseases such as malaria and plague. However, vectors can sometimes also be pests. For instance, the mosquito *Aedes aegypti* has gained notoriety because it transmits yellow fever and dengue viruses, but in both areas where it is, and is not, a vector it can be a pest because of the distress caused by its bites. Similarly, although house flies can transmit (mechanically) a variety of pathogens and parasites to humans and animals, they can be a greater problem because they can make outdoor recreational pursuits intolerable due to the large numbers landing on people, their food and drink.

Categories of pests and vectors

Major pests and vectors

What makes a major pest? Many things. When environmental conditions are most favourable arthropods can become very common and cause much annoyance or damage, or become important vectors. Major pests and vectors may be widespread or localized, or only important seasonally, such as during the summer or the rainy seasons when their populations are likely to peak.

The world's most dangerous and efficient malaria vector is most probably *Anopheles gambiae*, which is found in sub-Saharan Africa. What makes it so? Firstly, *Anopheles gambiae* is fairly common and is closely associated with man, breeding in aquatic habitats found around houses while adults commonly rest in houses before and after blood-feeding. More importantly adult females feed predominantly on humans in preference to cattle, thus there is a so-called high degree of mosquito–man contact. Adults are also *relatively* long-lived, which increases the chance that the developmental cycle the malaria parasites must undergo in the vector after their ingestion with a blood-meal, before the mosquito can infect a person with malarial parasites (sporozoites) while biting, is successfully completed. Typically the percentage of females that are infective, that is have malarial sporozoites in their salivary glands (called the sporozoite rate), is around 5%. In contrast a common mosquito in the Indian subcontinent, *Anopheles culicifacies*, actually bites cattle more often than man, the average life of adult females is slightly less than that of *A. gambiae*, and the sporozoite rate (% females infective) is usually 0.1% or less, yet in many areas it is the principal malaria vector. How is this possible? The answer is that local populations of *A. culicifacies* can be very large, so that people receive many more bites than Africans get from the more efficient vector, *A. gambiae*. Efficiency is replaced by numbers.

Theileriosis is an important tick-borne protozoal infection of cattle and sheep in Europe, North America and many tropical regions. It causes morbidity and loss of productivity in indigenous breeds, and severe, often fatal, disease in imported livestock breeds. One common form of the disease (caused by *Theileria parva*) in Africa is known as East Coast fever.

Every agricultural crop has its own list of pest problems. Although some pests are very polyphagous (e.g. armyworms), the secondary plant chemicals characteristic of botanical taxa deter the majority of potential herbivores and at the same time attract insects specializing on that taxon. Thus there is an 'allocation of grazing privileges' which means that each crop will have its own cohort of 'major pests'. As different crops are often characteristic of different continents, the concept of what are the major pests will not only differ by crop but also by geographical region (especially tropical, subtropical and temperate). Hill's (1975) lists of pests by crop for the tropics contain well over 300 different major pests of crops in the field and in store. A very restricted selection therefore has to be made here; we have selected a round half-dozen of non-migratory pests (migratory pests are discussed under a separate heading below) on the basis that any course on world agricultural entomology would not be credible without including these.

Although much international money is put into the control of migratory grasshoppers, i.e. locusts, local non-migratory grasshoppers are very often a much more serious problem for local farmers in the tropics, largely because they are equally voracious but occur regularly every crop season. They are very polyphagous and weed plants hold reservoir populations at all times.

Two hoppers figure in our list of major pests. The brown planthopper (BPH) of rice (*Nilapavarta lugens*) is a serious problem of the world's most important subsistence cereal, and is distributed wherever rice is grown in the tropics and subtropics. It has replaced the green rice leafhoppers (*Nephotettix* species) as the major hopper problem of rice with the spread of newer highly BPH susceptible rice varieties and the destruction of its natural enemies by insecticides, particularly the synthetic pyrethroids. Our other hopper is the maize leafhopper, *Cicadulina mbila*, again chosen because of the importance of maize as a staple cereal and the fact that this hopper transmits the crippling maize leaf streak virus. One of the problems with this hopper is that it can be found on a wide variety of wild grasses, and can spread from these over large distances.

The bollworms of cotton (especially *Helicoverpa armigera* of the Old World) have to figure in our list since they are the key pests of cotton, which is a very valuable crop and accounts for the largest proportion of insecticide use across the world. It is often said that if a new insecticide has no use on cotton, it is probably not worth marketing! The pest is polyphagous and is also an important pest of maize and legumes, as well as having many weed hosts. Although foliage is also eaten by the caterpillars, bollworms are mainly damaging to fruiting structures as internal feeders in cotton bolls, maize cobs and bean pods. The larvae are therefore concealed from pesticides and, anyway, all but the very youngest larvae seem to have a strong natural resistance to toxins.

We have to include the diamond-back moth (*Plutella xylostella*), since the high levels of resistance to most insecticides that overuse of the latter has caused in South-East Asia, coupled with the insecticides eliminating natural enemies, make this pest a research priority in developing sustainable pest management systems. The pest has many generations and a high fecundity in warm countries, and huge populations can develop quickly. The first stage larvae mine in the leaf, but then surface and the later instars strip the leaves of brassica crops in a short time, leaving only the main veins. In many countries, spraying against the pest is still done several times a week and it is pointless to include controls (unsprayed plants) in research experiments – such controls soon just become gaps in the field.

Fig. 1.1. A termite mound about 2 m high, Tanzania (H. F. van Emden).

Our last place on the list of six is allocated to termites. Many strip plants of leaf material to take back to their nests as the substrate for the fungus gardens that provide food for the colony. Also, the termite nests (termitaria) (Fig. 1.1) that many non-subterranean species build of excreta, soil and saliva form obstructions to mechanized agriculture and they can only be removed by explosives or the most robust earth-moving machines. Also, termites eat the wood of houses and farm buildings; the first shoes with stiletto heels sold in Africa quickly revealed the extent of termite damage as the owners fell through their house floors. Books are another source of food for termites, and also cockroaches, and need to be treated with insecticides against them in many parts of the world.

There are of course often seasonal differences in the importance of pests and vectors. For example, disease transmission by mosquitoes often occurs mainly, or only, in the rainy or monsoon seasons. However, when rice fields provide suitable breeding sites in the dry season, mosquito transmission can be more or less year round. Many crop pests in the tropics are also coupled to the rains for their emergence and appearance on the crop, which itself is sown to obtain maximum natural water. Yet a second season of the crop may be obtainable in the dry season, but with irrigation, and quite different pests may attack such crops compared with the rain-fed crop. Thus with cowpeas in

Nigeria, thrips may destroy the seedlings of the rain-fed crop, to be replaced by leafhoppers in the irrigated crop.

Body lice (*Pediculus humanus*) often become major pests, and sometimes also vectors, when people are crowded together, such as in refugee camps after disasters like earthquakes and floods. In these situations other pests, such as bedbugs may flourish and cause considerable distress because of their nightly biting. Biting lice (order Mallophaga) feed on skin fragments, feathers and hair and can be major pests of poultry, and to a lesser extent cattle and sheep, because of the irritation heavy infestations cause.

Minor pests and vectors

These can be arthropods whose populations are normally relatively small and often localized, but which under certain conditions can become pests or vectors; on the other hand populations can be quite large but they are inefficient vectors. A major pest in one area can be a minor one in another. Many minor pests of Europe (e.g. cereal leaf beetle, *Oulema melanopa*) have become major pests when introduced to other areas.

The tropical rat flea (*Xenopsylla cheopis*) is the major vector of plague, but a few other species may play a role in transmission. For example, the human flea (*Pulex irritans*) is now uncommon, but in the past used to be very common (this is the flea used in the now defunct flea circuses) and was probably a plague vector of some importance. Now the most common pest fleas of people are the cat (*Ctenocephalides felis*) and dog fleas (*C. canis*), which although preferring these pets will nevertheless attack their owners.

In most malarious regions there are minor anopheline vectors. These may be mosquitoes whose populations are small, or who prefer feeding on other hosts to people, or whose adult life expectancies are relatively short, or who are not very susceptible to infection with malaria parasites, or a combination of these and other biological factors. But despite this they may support some transmission of malaria, and are often regarded as secondary vectors.

Most crops have a long catalogue of minor pests. Hill's lists of pests for tropical pests have already been mentioned, and include far more minor than major pests. Just to give a few numbers, the lists (minor: major) for a few selected crops are as follows: apple (65:13), citrus (81:39), cotton (62:28), maize (44:32) and wheat (61:10) with rice going against the trend, having roughly equal numbers (48:49). So again we have to be very selective in giving examples. However, a good example is the apple blossom weevil, *Anthonomus pomorum*. This has a very short season of activity represented by one generation (egg to adult) a year all completed within apple blossoms. The rest of the year is spent as an adult sheltering in leaf litter and other debris close to the orchard. This pest is very rarely worth controlling; in most years one could argue it is actually beneficial in pruning excess blossoms which would otherwise develop into fruitlets requiring later thinning. Another example would be the large number (nearly a hundred) of scale insect species which have appeared on citrus as the result of the destruction of their natural enemies by insecticides aimed at other pests, but remain of minor importance. As a third example, we can point to one of the cereal aphids, the rose-grain aphid, *Metopolophium dirhodum,* which regularly appears on the lower wheat leaves and causes some damage, but rarely enough to warrant control. This is because, unlike the more serious grain aphid, *Sitobion avenae*, the rose-grain aphid does not colonize the flag leaf (damage to which has a big effect on yield) or the ear itself. However, unusual weather conditions (as occurred only in 1978 in our experience) can make it necessary to implement control measures.

Occasional pests and vectors

These are arthropods that are not normally pests or vectors, but may become so because of either explosive increases in their population size, or environmental changes that increase contact between them and plants, people or livestock. Examples are the dramatic population increases in mosquitoes following floods, or construction and engineering operations that create mosquito breeding sites and at the same time bring in workers who provide a readily available blood-source for hungry female mosquitoes. In 1977 wind-dispersed biting midges (*Culicoides imicola*), probably originating from Turkey, arrived in Cyprus and caused an epidemic of blue tongue, a viral disease primarily of sheep. The Australian bush fly (*Musca vetustissima*) is a troublesome pest of man and livestock because adults feed on sweat and mucus. Pest outbreaks often occur outside its normal distribution when flies get carried on warm northerly winds. In 1976, ladybirds became pests in southern England and appeared in large numbers biting people. The very hot year had temperatures lethal to the large aphid population which had developed earlier in the year and sustained a large population of ladybirds. When the aphids 'died of heat-stroke' the adult ladybirds dispersed looking for a new food source, but became concentrated into large clouds near the coast where the onshore winds prevented them going further.

When flooding or earthquakes devastate an area there are usually millions of animal carcasses, and sometimes human ones, such as in the large earthquake in Gujarat, India, in 2001. These carcasses provide fertile breeding grounds for 'filth flies' such as bluebottles (*Calliphora* species) and the resultant enormous increase in fly populations can be very troublesome as well as posing health risks. Flooding is also usually the cause of locust swarms. Locusts may exist for long periods in what is known as the 'solitary phase', dull coloured and in small numbers on the vegetation (often grasses) in the so-called 'outbreak areas' which are usually river valleys. When flooding occurs, the insects crowd onto what vegetation projects above the flood. Such crowding engenders a switch to the 'gregarious phase', in which the insects are much more brightly coloured and mobile, eventually taking off to swarm (see 'Migrant pests and vectors' below).

Staphylinid beetles of the genus *Paederus* are almost worldwide but are much commoner in tropical countries. They are not usually very numerous, but when their populations become large they can become very troublesome. Their haemolymph contains a vesicant fluid and when the beetles are accidentally crushed and rubbed on the skin they cause blisters; if they get in the eyes these become very inflamed. This is a good example of an occasional pest. (The more notorious meloid beetles, commonly known as 'Spanish fly' (*Lytta vesicatoria*), cause much more serious blisters.)

A French advisory entomologist, at the end of his career, published a list of occasional pests that he had encountered. These caused local problems for particular farmers for reasons specific to them – for example, a regular problem of weevils on carnations that were growing adjacent to a relatively uncommon weed related to carnations.

Migrant pests and vectors

The locust, probably the pest best known for its migrating swarms, evidences that our categories of pests are not mutually exclusive. Yet the sporadic occurrence of such swarms also makes the category of 'occasional pests' appropriate and the damage locusts can cause when they swarm would also qualify them as 'major pests'.

Most of the time, locusts live and breed harmlessly in wild vegetation, principally grasses, in their 'outbreak areas'. This term derives from the fact that it is from these areas, often river flood-plains, that the sporadic damaging outbreaks stem. In their outbreak areas, the locusts are in the 'solitary' phase. They are dull in colour, feed alone and have a relatively low rate of population increase. Outbreaks arise when the solitaries switch to the 'gregarious' phase. Gregarious locusts are brightly coloured, very active, crowding together and

Fig. 1.2. Part of a swarm of desert locusts in 1958 in Ethiopia. The entire swarm spread over 1040 km² (Courtesy of Natural Resources Institute, University of Greenwich).

breeding rapidly. The switch between phases is caused by an environmental event such as flooding or extreme drought, which forces the solitaries to crowd together on the little green vegetation remaining available. It is these gregarious flight-active locusts which swarm (Fig. 1.2) and, particularly on the African continent, this behaviour coincides with the development of the intertropical convergence zone of fast winds, which sweep the locusts to destinations across north-east Africa, India, South-East Asia and even as far as Australia. Here plagues then occur on the crops, the immature locusts (hoppers) produced marching like an army, devouring every green leaf or shoot in their path. After some months, the plague dies away (usually through starvation) and all that is left are solitaries in the outbreak areas hundreds of kilometres away.

There are annual predictable migrations of many insects, including pest species. These migrations can be over very large distances. For example, moths such as the silver-Y (*Plusia gamma*) whose larvae (cutworms) damage many crops, are typical of several species which breed in the Mediterranean and North Africa during the British winter, but migrate north annually and breed in the summer at more northern latitudes, while the vegetation in more southern climes has dried out in the summer heat. Of course, the migrations of insects are not analogous with those of birds, even if the results are similar. The insects are not navigating their journey. Evolution has selected for a flight behaviour timed to coincide with annual wind patterns that, in both directions, lead to the insects reaching destinations that enhance their survival. However, things can go wrong. There are many reports of locusts and moths landing in tens of thousands on the rigging of sailing ships hundreds of kilometres from land in the middle of oceans, and in 1958 vast swarms of diamond-back moth (*Plutella xylostella*) migrating from eastern Europe found themselves in the UK on the coast of East Anglia, instead of in North Africa.

Simuliid black flies (*Simulium damnosum*) have often reappeared in areas of West Africa where insecticidal dosing of their larval habitats, rivers, has eliminated them. These flies often disperse, on the wind, 200–300 km or more from their breeding places, and comprise mainly old females, many of which are infected with the nematode parasites that cause river blindness (onchocerciasis) in people. This clearly poses control and epidemiological problems.

Transportation of pests and vectors to new continents can also occur, not under the insects' own volition, but in association with the migration of humans and with their trading, and sometimes by animal migrations. For instance ticks, which may be important disease vectors feeding on birds, can be carried across continents when the birds migrate; and cattle can carry ticks, and other ectoparasites, considerable distances when driven along cattle trails.

In the eighteenth century slave ships racing their human cargo from West Africa to the Americas necessarily carried many casks of drinking water. It seems these often provided breeding places for the yellow fever mosquito, *Aedes aegypti,* and must have aided the spread of this mosquito into the New World.

Mosquitoes were unknown in Hawaii until the nineteenth century when in 1826 it is believed that a sailing ship introduced *Culex quinquefasciatus*, the mosquito vector of filariasis. In around 1860 *Anopheles gambiae*, the most efficient African malaria vector, was introduced from Africa or Madagascar into Mauritius. In this former malaria-free island there followed malaria outbreaks, including a disastrous one in 1867 which killed some 32 000 people.

Many of the insect problems of the new agriculture brought by Europeans who settled the Americas and Australia are European species which have made the same journey. One devastating example was the lucerne aphid (or spotted alfalfa aphid) (*Therioaphis trifolii*). It is believed that just one mother (aphids largely reproduce asexually) was all that was accidentally introduced into New Mexico in 1954, yet by 1956 the pest was causing damage of over US$ 40 million to the crop, and lucerne in California could hardly continue to be grown. The problem had an interesting solution that we relate in Chapter 13.

Another particularly serious problem in the USA of European origin is the European corn borer (*Ostrinia nubilalis*). This was introduced around 1908 and spread to 36 states by 1950, at which time losses were worth US$ 350 million. It is now such a major pest in America that it has been one of the first targets for genetically modified pest-resistant crops.

The oriental latrine-fly, *Chrysomya megacephala*, is a native of South-East Asia and Australasia where adults can be annoying pests and aid in the transmission of various enteric infections. In 1977 and 1978 they were suddenly found in West and South Africa, the Canary Islands, Peru and Brazil. It is assumed that increasing commercial trade, mainly shipping, aided their spread to these, and other countries, in the 1970s onwards.

In 1930 *Anopheles gambiae* was accidentally transported in a mail-plane or ship from Africa into Brazil, where it began to breed and flourish in the port of entry, spreading inland. In 1938 this mosquito caused malaria epidemics involving about 200 000 people and about 14 000 deaths. The enormous danger of the situation was realized and a semi-military-style campaign was launched which in just two years eradicated the mosquito from the country, at a cost of US$ 42 million. It has never reappeared. In 1942 this same species invaded Egypt, possibly through river traffic from the Sudan, but was eradicated in 1945.

The mosquito *Aedes albopictus* is a native of South-East Asia where it commonly breeds in discarded motor vehicle tyres. In common with other *Aedes* species eggs can tolerate months of desiccation and then hatch when soaked by rainwater. In 1979 *Aedes albopictus* suddenly appeared in Albania and in 1985 in Texas, USA, where it is now breeding in at least 28 US states. It has also been introduced into Mexico, South America, Fiji, Italy, Nigeria and a few other countries. It was brought to these countries as dry viable eggs in used tyres exported for retreading. Apart from becoming a pest mosquito in urban and semi-urban areas of the USA, there is concern that in some countries it could become a vector, because in its native South-East Asia it transmits dengue virus.

The so-called jigger or chigoe flea (*Tunga penetrans*) is a native of Central and South America. The adult female of this flea buries herself in the soft skin of the feet, especially between the toes. The site of infestation often becomes infected with pathogens and in extreme cases gangrene can occur. Can this flea be regarded as a migrant pest, albeit a rare one? The question is asked because infected people carried the flea, in their feet, to Africa towards the end of the nineteenth century where it is now a common pest. It is also occasionally recorded in the Indian subcontinent, having been brought there by infected people returning to India. We think the answer is yes.

More recently, in 1999 and 2000, New York experienced outbreaks of West Nile virus, the first time this virus had appeared in the Americas. It caused headline news in the USA, with some degree of panic that a tropical virus had hit metropolitan New York. It seems probable that infected mosquitoes from Africa, Europe or Asia were transported by aircraft into the USA, although it is possible that virus-infected birds may have been responsible. By 2003, West Nile virus had been recorded in 44 states of the USA, as well as in Canada and Mexico, a truly remarkable rapid spread.

Potential pests and vectors

That problems have arisen as a result of environmental changes or man's activities indicates that there are probably many such potential pests and vectors among the insects we currently consider as 'economically neutral'. The likelihood of 'global warming' has stimulated a large amount of research to predict expected changes in the pest and vector spectrum in different parts of the world (see Chapter 2). The Russian wheat aphid, *Diuraphis noxia*, has greatly increased its distribution in recent years with the invasion of both North and South America. Much computer modelling has sought to establish whether this species could become a pest in other cereal-growing areas, particularly northern Europe and Australia. If global warming occurs, then several tropical diseases, including malaria, could spread to formerly cooler areas (see Chapter 2). There is a reservoir of infection in the UK, mainly in people from the Indian subcontinent, but as these people live mainly in urban areas where there are few, if any, anophelines there is little chance of malaria transmission. If they moved to rural areas then that is a different scenario. During the First World War, invalid soldiers returning from malarious areas were stationed in rural Kent and local anophelines transmitted several hundred malaria cases.

Has the Colorado beetle the potential to be a pest in the UK? It is occasionally reported, triggering application of insecticide where it has appeared. Yet, although each year holiday-makers bring the attractive beetles back to the UK, it has never established itself.

Florida is not a wine-producing area, in spite of its favourable climate, because Pierce's disease (a bacterium), widespread in many perennial crops and ornamental plants can destroy the vines. The disease is vectored by a xylem-feeding leafhopper, the glassy-winged sharpshooter (*Homalodisca coagulata*). This has always been seen as a potential pest for the vineyards of California, a state where the bacterial disease has been present since the nineteenth century, but has no efficient vector. The appearance of the sharpshooter in California in 2000 has caused considerable panic.

Agricultural practices and disease

The development of agriculture has had a dramatic effect on pest problems. The appearance of large monocultures of single crops, and the agronomic practices carried out there (including the advent of insecticides) has enabled many species of insect to multiply to pest proportions. All this is covered in other chapters (especially Chapter 2). However, the other effect of agriculture has been to alter landscapes and socio-economic conditions, and it is these large-scale changes which have had such an impact on humans through the prevalence of diseases of people and livestock. It is this aspect which is covered in the following section.

The world population is predicted to grow from 5.8 billion to an estimated 10 billion by the year 2050, with alarmingly 90% of this increase in developing countries. This expanding population will require more food, but this needs to be grown in the developing countries, not imported from richer countries.

About 30% of the earth's surface is classified as arid or semi-arid, but some 600 million people struggle to live in this inhospitable environment. Irrigation is one method enabling more food to be grown in such areas. An example of successful irrigation in arid regions is Israel. Not only is sufficient food produced for home consumption but there is surplus for export; unfortunately not many countries operate such efficient irrigation. But irrigation and other agricultural practices can have deleterious effects on people's health. Some examples follow.

Irrigation

Of all crops irrigated rice provides the worst disease scenario mainly because large areas are flooded for long periods. Presently there are about 290 million hectares of irrigated land, of which rice is grown on about 145 million hectares.

Unfortunately, flooded rice fields (Fig. 1.3) can generate phenomenal numbers of mosquitoes including anophelines. In Kenya the numbers of the malaria vector, *Anopheles gambiae*, biting people living on, or near, an irrigation project (at Ahero, near Kisumu) is about 70 times greater than those biting villagers a few kilometres distant. Such increased vector populations give rise to increased malaria transmission. On the Cuckorova plain near Adana town, Turkey, irrigation schemes were accompanied by an influx of a migrant labour force, many

Fig. 1.3. Terraced rice fields in Sri Lanka, typical larval habitats of mosquitoes such as *Anopheles* species and *Culex tritaeniorhynchus*, vectors respectively of malaria and Japanese encephalitis (M. W. Service).

of whom were infected with malaria. Large populations of *Anopheles sacharovi* breeding in the flooded fields resulted in an explosive increase in malaria.

If there is double cropping of rice to increase productivity there can be two seasonal malaria peaks instead of just one in the rainy season, which can result in more or less all year round disease transmission. More than 40 arboviral (arbovirus = **ar**thropod-**bo**rne **virus**) infections are associated with rice growing; by far the most important is Japanese encephalitis which is a widespread human disease in Asia and is transmitted by the mosquito *Culex tritaeniorhynchus*. In 1978 the Sri Lanka government initiated the Accelerated Mahaweli Irrigation Scheme, for growing mainly rice, on about 127 000 ha of land. Not surprisingly rice growing caused dramatic increases in *C. tritaeniorhynchus*. Over a period of seven years 150 000 families were resettled on the scheme, and in one area farmers were advised by agricultural experts to keep pigs to supplement their income. Pigs are regarded as amplifying hosts for Japanese encephalitis, that is when mosquitoes bite them they develop a very high virus titre in their blood. So the mixture of pigs and mosquitoes proved explosive, and from 1985 to 1989 there were outbreaks of Japanese encephalitis. This serves as a warning how the lack of dialogue between agriculturists and health workers can lead to problems.

Although *Culex quinquefasciatus*, a vector of bancroftian filariasis, does not breed in rice fields because it prefers organically polluted waters, its population can nevertheless increase dramatically in adjacent areas where tenant farmers and their families live. This is because there is often a proliferation of organically polluted waters, such as flooded pit latrines and blocked drains.

Increased human populations near rice fields and the building of numerous granaries attract rodents, such as rats, which may be infested with tropical rat fleas (*Xenopsylla* species), vectors of bubonic plague. Irrigation schemes, in fact, often cause widespread ecological and environmental changes, such as increases in bird populations, especially aquatic ones, which can be reservoir hosts of arboviruses transmitted to man and domesticated animals by ticks and mosquitoes. There may also be a reduction in livestock, especially cattle, and wild animals.

Deforestation

A major impact on land use in sub-Saharan Africa is from animal trypanosomiasis, which precludes large areas being economically farmed because of the presence of tsetse fly vectors. This results in cattle being concentrated in dry areas where tsetse flies cannot survive, and this in turn gives rise to overstocking and land degradation, and cattle of inferior quality. When farmers clear forest and scrub land to plant crops, such as cassava, yams, cotton and groundnuts, a more open habitat is created and this is unsuitable for tsetse flies. Perhaps even more importantly is the resultant reduction in wild animals upon which the flies feed, which also results in a decline in tsetse fly populations.

Deforestation can also have an impact on shade-loving anopheline mosquitoes. There are well documented cases in Nepal, Sri Lanka, Malaysia and Thailand where clearing land for farming has eliminated relatively inefficient malaria vectors, and replaced them with more efficient sun-loving ones, and an accompanying increase in malaria transmission. In Trinidad in the 1930s and 1940s forested areas were cleared for cocoa plantations. Shade trees were then planted to protect the young trees from excessive sun; these became colonized by epiphytic bromeliads (Fig. 1.4), which in turn were colonized by *Anopheles bellator,* an important malaria vector. As a consequence there were malaria outbreaks in the cocoa estates.

Farm mechanization

The increase in rice production that has occurred in developing countries has been achieved in part by mechanization, which is proceeding fastest in Asia,

Fig. 1.4.   Epiphytic bromeliads growing on trees in Trinidad. Their leaf axils when containing water are larval habitats of the malaria vector, *Anopheles bellator* (M. W. Service).

particularly in China, Korea, Thailand and Malaysia. Farm mechanization, such as use of tractors, can lead to (i) more crop cycles a year, (ii) increase in farm hectarage, (iii) changes in land use, (iv) cultivation of marginal lands, (v) increased use of fertilisers and pesticides and (vi) reduction in livestock. Many of these changes can have health repercussions, obviously good if more food is grown, but there can also be adverse effects. For instance, additional crop cycles per year and cultivation of more land can increase and extend mosquito breeding more or less throughout the year. Greater usage of pesticides to combat crop pests can induce insecticide resistance in medical vectors (see p. 194). However, one of the more interesting and possibly least-expected consequences is how a reduction in livestock can affect mosquito production and disease transmission. Cattle hoof-prints fill up with water and provide ideal larval habitats for several anopheline mosquitoes; furthermore construction of ponds to supply drinking water for cattle can provide mosquito breeding places during dry seasons, and thus generate large vector populations when they would normally be negligible. Most haematophagous arthropods which are pests of man or disease vectors will also feed on animals, including livestock. Now, it has been estimated that in Pakistan and Bangladesh the introduction of a tractor replaces 2.0–2.5 bullocks, originally needed for ploughing. If there is a reduction in cattle then mosquitoes will likely feed increasingly on people, and this can lead to disease outbreaks. One of the more interesting and convincing examples of such ecological change occurred in Guyana. Prior to the 1960s malaria was transmitted by *Anopheles darlingi*. Because adults of this vector rest inside houses an eradication campaign based on spraying the interior of houses with DDT virtually eliminated this vector and malaria was eradicated along a coastal area, including the Demerara river estuary. *Anopheles aquasalis* was also common but, because adults tended to rest outside, house-spraying had little effect on its population size. This did not matter because as it fed on livestock it was not considered a vector. After malaria was eradicated the human population increased and as people became more prosperous they sold their cattle, concentrated on growing rice and replaced oxen with tractors for ploughing. They bought trucks for transportation purposes instead of donkeys and mules. Because of this deficit in livestock the originally zoophagic *A. aquasalis* switched to feeding on people. Arrival of malaria-infected itinerant workers reintroduced malaria parasites into the community. As a consequence malaria returned to the Demerara river estuary 16 years after it had been eradicated. There are other suspected, but less well-documented, cases of reduction in livestock resulting in increased disease transmission.

Beneficial impacts of insects

This aspect of insects should be mentioned to balance the emphasis in this book on their pestilential properties. However, since this is a book on controlling the pestilential species, the beneficial impacts can only be given a brief mention.

Scavenging

Most people would probably put pollination or honey at the top of the list, but in reality the most vital insect activity for human welfare is probably the recycling of plant and animal material. Termites, of which some species are so damaging to crops and buildings (see earlier), are on balance probably beneficial as an Order of insects. Many species of termites break down dead plant material which would otherwise accumulate on the soil surface; they have been called 'the earthworms of the tropics'. Insects are also important in burying and decomposing corpses of small animals and birds and excreta. When, in 1788, European cattle were brought to Australia their cowpats accumulated since the local dung insects were adapted only to the excreta of marsupials. With more than 300 million dung pats being deposited on the surface of Australia by the cattle every day, and remaining unprocessed, much of the rangeland was rendered useless for grazing. Moreover the irritating bush-flies (*Musca vetustissima*), against which cartoon Australians wear broad-brimmed hats with corks dangling from the edge, bred in the dung and greatly increased in numbers. However, government scientists successfully controlled the dung problem biologically by introducing several species of African dung beetles to Australia.

Pollination

This is probably the second most important benefit of insects to man. Although many plants do not rely on insects for pollination (e.g. wind pollination), many do so, either largely or entirely. The flowers of many plants are clearly adapted to insect pollination, with attractant colours and scents. Some flowers are designed for insect access; one has only to watch a bumble bee (*Bombus* species) landing on the lower lip of a snapdragon flower (*Antirrhinum majus*) to see how the flower then opens to allow the insect access to nectar while at the same time contacting it with its anthers to leave pollen on the back of its visitor.

Many crops are grown because they yield fruit or seeds as the result of pollination; bees are thus important and effective visitors in agriculture and horticulture. Although bad weather and high winds deter them, 100 bees can set a commercial crop of one hectare of apples in five hours. Growers may rent beehives to ensure pollination, and cardboard bumble bee nests are sold to glasshouse growers for crops such as tomato which require pollination.

However, many wild insects visit and pollinate crops as well as bees, and may be the chief pollinators of crops such as cherries, which flower early at a time and under temperature conditions when bees may be largely inactive.

Some crop plants have associations with other insects for their pollination requirements. Cocoa in the tropics is pollinated almost entirely by non-biting nectar-feeding midges (*Forcipomyia* species), but the most specialized association is probably that of the Smyrna fig with its fig wasp (*Blastophaga psenes*). If a fig flower is not pollinated, the seeds do not form and flavour is impaired. The Smyrna fig produces only female flowers, so a pollen-producing variety is also needed. Pollination is carried out by the female of the tiny fig wasp. Eggs are laid in the flowers of the pollen-producer and the larvae develop in small galls at the base of the flowers. The male wasps are wingless, and after fertilization, escape to seek new flowers for oviposition, entering by the small hole which is the only access to the flower. The ovaries of the Smyrna fig are deep-seated, so oviposition by the wasp rarely occurs, but pollen is left behind in the wasp's attempts to reach the ovaries.

Insect natural enemies

Carnivorous insects may use pests and vectors as their food and so reduce or eliminate the need to control their populations. This role in the population dynamics of insects is explored in more detail in Chapter 2, and how carnivorous insects may be intentionally manipulated to control pests or vectors is the subject of 'Biological Control' (Chapter 7). Therefore only the heading is given here, to remind the reader of this important aspect of the benefits man derives from insects, However, one area not covered elsewhere in this book is the biological control of weeds by herbivorous insects. This is a very active area of biological control, for weeds need reducing rather than controlling to very low levels, so biological control of weeds is more likely to succeed than biological control of insects. The classic landmark for biological control of weeds was the spectacular control of the prickly pear cactus (*Opuntia*) by the South American moth, *Cactoblastis cactorum*. There have been many other success stories since.

Insect products

Man derives a number of valuable products from insects, and may even farm and protect them for this purpose.

Honey from bees is probably the oldest such product – since bears rob wild hives for their honey, it is likely that man has also done so ever since the dawn of the race, and that bees were the first insects to be domesticated by man. Bees gather nectar from flowers and process it with enzymes in the saliva to feed the brood. The fanning of their wings moves the air in the hive and concentrates the solution. Man then can collect and extract the honey, capitalizing on the industry of the worker bees, who have had to make perhaps 100 000 journeys from the hive for every kilogram of honey the beekeeper extracts.

Modern hives are far removed from the hives constructed by wild bees in hollow logs, for example. Today hives consist of prepared honeycombs provided in boxes called 'supers', which can be stacked on top of each other as the season progresses and the bees need more space for their combs. In some parts of the world, e.g. Australia, beekeeping is a serious industry compared with how it is done in countries such as Britain. Scouts on motorcycles seek out likely sites of good later nectar flow in the forests, trying to shake off 'spies' from their competitors, and establish a safe site for juggernaut lorries later to deliver the hives and supers.

However, beekeeping provides a number of valuable products in addition to honey. Beeswax is a natural secretion of the worker bees, used to construct the comb and cap the cells containing the honey. Beeswax is extracted, and much is used to make 'foundation', the honeycomb in the supers. Beeswax is also used for candles, particularly church candles, cosmetics, wax polishes, crayons etc. There are also industrial uses in lubrication, lithography and electrics. The bees also collect tree gums and resins to make propolis (a brown sticky glue) to fill cracks and rough surfaces in their hives; this material has antibiotic properties. Finally from the hive, royal jelly is obtained and sold in various forms such as inclusion in face creams with claims that it will prolong human life. Royal jelly is a special food fed to larvae destined to become queens, and it is true that they will live for several years in contrast to a worker bee's six weeks. Although there is therefore no doubt that royal jelly possesses remarkable qualities, there is no evidence it is beneficial to humans. Those selling it fail to point out that what it achieves in bees is to maintain the queen as an egg-laying machine; most men and some women might not wish to be endowed with this characteristic.

The silkworm moth (*Bombyx mori*) is another insect which has long been domesticated by man, and remains the only natural source of silk. Silkworms have been reared for silk by man for nearly 4000 years, each caterpillar converting mulberry leaves into a single silken thread a kilometre long in the form of the cocoon within which the caterpillar pupates. About 6000 cocoons are needed for a kilogram of silk, with the insect within the cocoon first killed by heat to prevent it cutting its way out as an adult and so damaging the silk. Other beneficial insects include those that supply us with shellac, used in French polish as well as for many other purposes, and which is produced by the scale of the scale insect *Laccifer lacca*. The dried bodies of another tiny plant louse, the cactus mealybug *Dactylopius coccus*, are pulverised to obtain the food colouring agent cochineal (it takes about 200 insects to make a gram of cochineal).

Plant galls made by insects contain unusual pigments and other chemicals which are extracted as inks and medicines, particularly in the Orient. Recently a problem arose in the south-west of China, where there was a decrease in the aphid population on the tree *Rhus*, which carried the valuable aphid-induced galls. It is rare for a shortage of aphids to be a cause for complaint!

Royal jelly of bees and gall pigments are only some of the medicinal uses of insects. When maggots of greenbottle (*Lucilia*) and bluebottle (*Calliphora*) flies invade wounds they devour necrotic tissues and pus. Such flies were used in the pre-antibiotic era by surgeons, such as in the Napoleonic wars (1796–1814), to cleanse septic battle wounds. Despite an initial disgust at this practice several hospitals in the USA, the UK and in other parts of Europe are employing maggot therapy, that is using specially reared sterile maggots to clean up leg ulcers! Moreover, an accidental observation during the First World War led to the discovery of allantoin, which is secreted by maggots and has antibacterial properties. It has now been commercialized. Recognition of these antibiotic properties came before the discovery of penicillin. Cantharidin from blister beetles, once prized only as an aphrodisiac, has also been used as a treatment of the urinary system.

Insects as human food

In countries like the UK, eating insects is something eccentrics do on television to entertain us. It was also a temporary fashion in the 1970s, when an entrepreneur imported tins of insects from the Far East as novelty foods. For a time, it was possible to find tins of fried locusts, stewed caterpillars, bumble bees in syrup and chocolate-covered ants on our grocery shelves.

In contrast, insects have always been, and remain, an important source of food for many peoples, particularly in the tropics where insects attain a large size and become a worthwhile mouthful. Locusts, large caterpillars (e.g. the 'witchetty' grub of swift moths (Hepialidae)), termites or beetle grubs are commonly eaten fried, but it is not unusual to find insects eaten raw, such as the sweet swollen honey-pot ants which act as food stores in the nests of some ant species. The vast swarms of chaoborid and chironomid flies that arise from East African lakes are boiled and made into cakes. Often insects are an important source of food for poor local people; entomologists who have successfully controlled caterpillar pests have often finished up as very unpopular with local tribesmen (see Chapter 13). Aristotle reports that, although male cicadas are tastier when young, with females it is better to wait until they are full of eggs.

In Mexico, special sheets of cloth are hung under water to encourage aquatic bugs to lay their large eggs thereon. The eggs are then harvested and dried, and sold as an ingredient for cakes.

In the Near East, the honeydew of aphids and scale insects is collected in tens of thousands of kilograms and sold as a sweet confection. In St Mark's gospel, John the Baptist is reported as existing for some time on 'locusts and wild honey'.

Insects as food for birds and fish

Insects are a major source of food (probably well over half) for very many birds, including the chicks of game birds. The use of herbicides and insecticides in cereal fields reduced the survival of the grey partridge (*Perdix perdix*) in the UK because the number and diversity of insect food for the chicks at the field margins was greatly reduced. The cost in terms of lost shooting income was sufficiently large to persuade some farmers to allow flowering weeds and insects to flourish at the edge of the field by turning off the outer spray boom when spraying at the field edge (see Chapter 7).

That anglers use artificial insects ('flies') to lure fish onto the hook is evidence enough that fish take insects. In Illinois in the USA, the dependence of freshwater fish on insect food has been estimated at 40%.

Some other uses of insects

Luciferin from glow-worm beetles (Lampyridae) is used in high-tech equipment for measuring microbial activity in the soil, including its use by agrochemical companies monitoring possible side-effects of new insecticides. This is just one of many uses of insects in scientific study often in unrelated areas. After all, modern genetics owes a great deal to just one insect, the fruit or vinegar fly, *Drosophila melanogaster.*

Insect life, especially larvae of chironomids and simuliids, in streams at various distances downstream from factories is routinely monitored as a measure of industrial river pollution.

Insects are often beautiful and contribute greatly to the enjoyment by humans of their environment. Here butterflies obviously take pride of place, but there are also many other beautiful insects large enough to be noticed, including dragonflies and large beetles (Fig. 1.5). In the tropics, there are many large and remarkable insects such as the stick and leaf insects and the praying mantis. Jungle and butterfly houses are now very popular tourist attractions.

Insects have always been the subject of a hobby industry. The butterfly houses sell their dead specimens, and there are other sources of specimens for purchase as dead specimens or for rearing by collectors of different insect Orders. Linked with this is the sale of collecting and storage equipment, identification books etc.

Unfortunately, some humans have put a price on the heads of some of the most spectacular creatures and have collected particularly large tropical butterflies for pictures, brooches and the decoration of hats. Some countries

Fig. 1.5. Goliath beetle (*Goliathus cacicus*, Scarabaeidae), about 15 cm long (courtesy of www.thais.it).

such as Malaysia and Brazil have been especially despoiled to the point where the collection of insects for commercial profit has been made illegal. Sadly, the main effect of the legislation has merely been that we now call the collectors by a different name, smugglers.

Finally, insects can help solve murders! The succession of necrophagous insects in corpses can often help determine the time, and sometimes even the place, of death. This is part of the science of forensic entomology.

2

The causes of pest and vectored disease outbreaks

Introduction

As pointed out in Chapter 1, a commonly quoted statistic is that, without insect pests in the field and store, world food production could be increased by about a half. As this estimate represents the loss despite current control measures, it would clearly be catastrophic for mankind if control of insect pests were not attempted or should fail. The efficient control of certain disease vectors would save very many human lives, and then even more additional food would be needed to feed the hungry, even more rapidly expanding population. Thus more effective vector control would have the knock-on effect of creating a demand for even more effective pest control!

The pest problem

Obviously each insect individual has a fairly small food requirement. For example, a greenfly (aphid) is unlikely to extract more than about 0.5 cm^3 of sap from a plant in its lifespan, and even a voracious caterpillar is likely to consume only 50 g of the fresh weight of its host plant.

Although some pests, such as those that spoil a whole apple with one small blemish or those that vector plant diseases, can cause damage even at low populations, the damage done by most crop pests results from the enormous numbers in which they occur. There may often be 25 million insects per hectare of soil and 25 000 in flight over a hectare, compared with a human density over the dry land of the earth of about 0.2 per hectare. Obviously, not all these insects are eroding man's food supply; however, numbers of just a single pest species per hectare of crop will often be comparable with such figures. One hectare of oats may harbour 22 million frit fly (*Oscinella frit*) larvae and in Chapter 1 we quoted 200 million for the number of black bean aphids (*Aphis fabae*) which might be found on a hectare of sugar beet. Both these infestation figures represent a rapid multiplication from relatively few initial immigrants, and indeed most insect species, which cause pest problems, have amazing reproductive powers. This also applies to many of the vector species. In the tropics mosquitoes can increase explosively in numbers during the rainy season, maintain high population levels for a considerable time and then gradually decrease in numbers. High population densities of simuliid black flies and mosquitoes have been blamed for reducing both weight gain in cattle and milk-yield.

The statistics that have been calculated for the populations which insects could attain, if such explosive powers of reproduction were maintained for long periods, would not seem out of place in a science fiction novel. Thus, cabbage aphids (*Brevicoryne brassicae*) have the potential to produce a new generation every two weeks, with 50 young per female in each generation. This can be expressed more dramatically as the potential in one year of one aphid mother to produce offspring weighing 250 million tons, encircling the equator nose to tail a million times. Equally startling is the notion that in one year a pair of house flies (*Musca domestica*) could cover the earth to a depth of 15 m with their offspring (200 000 million individuals). Finally, two cabbage white butterflies (*Pieris rapae*) in one year could, with their offspring sitting with their wings closed, cover the surface of Australia but also building a tower rising into the stratosphere and beyond at the speed of light!

Such statistics are of course wildly unreal because they assume not only maximum fecundity sustained regardless of numbers, but also mortality, apart from death through old age, again remaining zero indefinitely.

Factors affecting the abundance of insects

In the real world, the balance between births and deaths, neither of which remains constant, determines the rate of population change. The cabbage whites in the above example would have starved to death long before covering the surface of Australia, making it clear that mortality, at least, is influenced by population density in the form of increased competition between individuals (intraspecific competition) for the remaining food resources.

It is inevitable that there has to be a resource ceiling to population growth, giving any area of habitat a constrained 'carrying capacity'. Forest caterpillars in

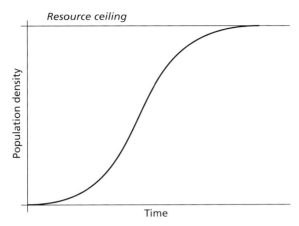

Fig. 2.1. Lotka–Volterra sigmoid curve of population growth against time.

the naturally regenerating forests of central Europe cause increasing defoliation year by year until the trees are killed and mass caterpillar starvation occurs. Both trees and caterpillars then begin again on something like a 50-year cycle. However, such a dramatic and late kicking-in of density-dependent mortality leading to what is known as 'scramble competition' seems to be avoided by the majority of animals.

Most animals regulate their own populations to minimize overcrowding by replacing scramble competition with some form of contest competition within the species. If we rear a single species in a uniform environment (a stored products pest in a large jar of flour would be a good example), the population rises following the pattern shown in Fig. 2.1, the famous Lotka–Volterra curve developed as early as 1925. This curve shows that the maximum population growth rate with which the curve starts lasts a very short time, and continues to slow progressively until the population approaches the resource ceiling at a snail's pace.

What is happening is that the population is adjusting the balance of births and deaths to enable the population to survive as long as possible. If we were to provide an escape route from our jar of flour, the population would survive even longer as we would note an increase in the number of adults emigrating to seek new habitats as the population increased. So, not only is the population reducing its reproductive rate early in response to crowding, but is also increasing population reduction equivalent to death by emigration. A good example of such reductions of increase rate in response to crowding is seen in the blood-sucking triatomine bugs (e.g. *Triatoma infestans*). Increases within houses of numbers of these bugs leads to a reduction both in the proportion of bugs successfully feeding and in the size of the blood-meals.

The result is to prolong the life-cycle, reduce fecundity and often stimulate flight out of the house. There is no increase in mortality, but the population is adjusted to a lower carrying capacity.

With some mosquitoes there is evidence that when larval aquatic habitats are overcrowded this results in the production of adults whose behaviour is modified in such a way that they tend to fly upwards and get dispersed away from the area by the wind, thus relieving competition for limited resources. Aphids provide another example. They put their effort into reproduction when food is plentiful and avoid the expense of making wing muscles by remaining as wingless adults. However, when they become more crowded, contacts between individuals increase, and an increasing proportion of winged emigrants are produced. Other examples of a feedback into the system include the host defensive reactions of pecking by birds, and tail switching and ear flicking of cattle, all of which increase with the more flies attempting to feed. These actions will prevent some flies from feeding while others will take smaller quantities of food which will result in fewer eggs being produced. In the 1970s a small scale genetic control trial in Kenya against the mosquito *Aedes aegypti* resulted in 64% of the eggs laid in village water-storage pots being sterile. But this did not cause any reduction in the numbers of adult mosquitoes. Why? Because with fewer mosquito larvae in the pots they all obtained sufficient food to complete their development; previously there was fierce intraspecific (i.e. scramble) competition for a limited food supply and this resulted in many fewer achieving adulthood.

Territorial behaviour is common among insects, and is a form of contest competition called 'conventional competition'. Here 'conventional' is used in the sense of a convention as a 'token'; i.e. the animals compete for a token resource with a lower carrying capacity than the real resource. Thus territories are often used to represent the food resource, with territory availability limiting population density well below the food resource ceiling. An example is found with caterpillars. These are normally considered vegetarians, but those of many species can be cannibalistic with fights occurring with increasing frequency as territories become occupied. This is illustrated by the cotton bollworm, *Helicoverpa armigera,* which has a very different number of territories on two of its crop host plants, cotton and maize. Cotton has up to 90 territories (the bolls), so has a large carrying capacity for bollworms before they are likely to meet and fight. Bollworm can thus have large populations on cotton, causing severe economic damage. By contrast, bollworm on maize competes for only a few territories, the maize cobs. Cannibalism then occurs within the cobs, and there will be only one survivor per cob, unlikely to damage a high proportion of the maize seeds. The world's largest mosquitoes belong to the genus *Toxorhynchites*; the adults are large, colourful and non-blood sucking. Their larvae occur in container habitats such as water-filled tree holes and are voracious predators, but they are also part-time cannibals, although not all victims they kill are consumed.

Such behaviours, coupled with contest competition resulting in adjustments of fecundity to changes in density, not only reduce the dangers of overcrowding, but equally enable populations rapidly to escape from dangerously low densities (undercrowding), when the chances of the sexes meeting are so low that extinction may be inevitable. The North American passenger pigeon (*Ectopistes migratorius*), which once formed flocks of millions of birds, was hunted to extinction in the wild by 1900. The last specimen (in a zoo) died in 1914. Similarly the Mauritian dodo (*Didus ineptus*) died out on the island in 1681, following hunting by European settlers; it had not encountered predators before and made no effort to escape when approached by people with guns! The American bison (*Bison bison*), which once roamed the plains in millions, was reduced to less than a thousand animals by 1885 and to virtual extinction by 1900, though breeding programmes have now raised numbers in the USA to about 200 000. The animals in these examples did not became extinct because man killed the last few, but because their populations became irretrievably undercrowded.

Regulation to keep the population fluctuating within intermediate limits is therefore something all populations seek to achieve, since there is always the danger that 'events beyond their control' may suddenly move the population level dangerously upwards or downwards. Such changes are caused by density-independent events, the impact of which is proportionally the same at all population densities. Such events include weather; for example, a sudden frost may kill the entire population, whether few or many – similarly a sudden rise in temperature will speed up generation times, again regardless of population size. Sudden events also have density-independent effects on mosquito populations. Here desiccation of larval habitats often causes the greatest population loss, while heavy rainfall is usually accompanied by greatly increased numbers. This demonstrates nicely how weather events can disturb natural equilibria, the antithesis of population regulation. Insecticides are another example of a disturbing (though unnatural) density-independent factor. Although its advent is density-dependent in the sense that insecticides are only likely to be applied when populations have reached a certain size, the proportion killed will be very large and not related to the density of the population.

Figure 2.2 presents a classification of natural environmental factors in relation to the dependence of their impact on the density of a population.

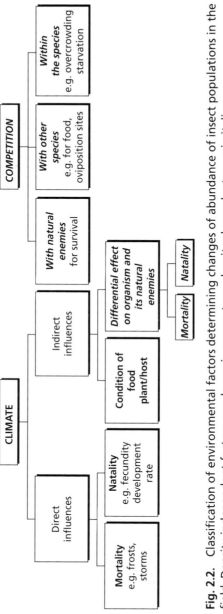

Fig. 2.2. Classification of environmental factors determining changes of abundance of insect populations in the field. Density-independent factors are shown in roman type, density-dependent ones in italics.

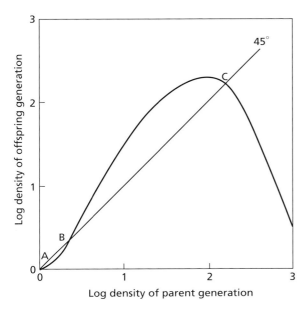

Fig. 2.3. Plot of log offspring density against log parent density.
A, extinction; B, point on 45° line where 'undercrowding' begins further to
left; C, point on 45° line where 'overcrowding' begins further to right.

It is thus clear that population growth rate will rarely be one – the 'zero growth' rate where the population remains constant since the offspring generation is exactly the same density as the parental one. This can be illustrated by plotting the straight line of a zero growth rate for the density of the offspring generation against the density of the previous (parent) generation (Fig. 2.3). We can already mark three points on this straight line. A – Zero parents are bound to have zero offspring! B – As undercrowding will lead to fewer offspring than parents, there must be a point on the line where growth rate moves from less than one to greater than one (the normal state). C – At extremes of overcrowding the population must fall, therefore there has to be another point on the straight line where the growth rate reverts to less than one. Now, given that at undercrowding and overcrowding the population growth rate is less than one, but in excess of one at other times, it is straightforward to complete the curve of offspring generation against parental generation density (Fig. 2.3).

The next stage is to model the change to this curve following from the introduction of a predator, acting in a positively dependent way so that the percentage mortality it causes increases with increasing prey density. But first we need to ask why a predator should impact in a density-dependent way. The

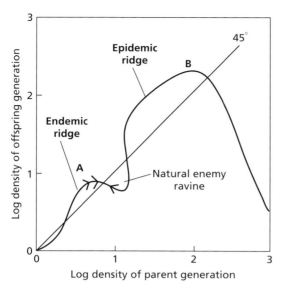

Fig. 2.4. Plot of log offspring density against log parent density with introduction of density-dependent mortality from biological control. A, regulation by the biological control; B, regulation by intraspecific competition. The 'Endemic ridge' and the 'Epidemic ridge' identified on Fig. 2.5 are indicated.

answer has two components; one is numerical and the other functional. The numerical response is that a greater density of prey will attract more predators, and they will be more inclined to stay and breed. Thus the numerical response affects how many predators there will be. The functional response relates to a different concept – how many prey each individual predator will kill. This is a function of how much time a predator will have to spend searching for prey between feeding bouts, and how long it spends with each meal. As prey density increases, predators do not have to spend as much time searching, but also they often spend less time handling each prey in that they just 'eat the best bits' and then move on.

So the impact of a positively density-dependent predator on our graph is increasingly to bend the line to a lower population growth rate (Fig 2.4, point A). This continues to a point where the impact is relaxed and growth rate increases again (single arrow on figure) until further increase in growth rate is again prevented by the predator (double arrow on figure).

This is the action of straightforward positive density-dependence. Sometimes density-dependence is delayed by a generation or year (delayed density-dependence). This can happen when predators do not prevent a year of high prey density, yet many survive to attack the prey the following year and prevent its increase. As many predators would then starve, few survive to the following year, allowing large prey populations again to develop. Such delayed density-dependence can achieve regulation in the longer term, but with much larger fluctuations around the equilibrium population than if density-dependence is not delayed.

There is also inverse density-dependence, when the prey are so abundant and breed so fast that predators just cannot keep up or may even become satiated, all leading to a reduced proportional impact as prey density rises still further. This escape from regulation by predators may occur if there is a more rapid increase in the population than the predators can meet by stepping up their numerical and functional responses. The prey population will then rise until intraspecific regulation responses take over to reduce population growth (Fig. 2.4, point B). Southwood (1975) makes the analogy between the graph and a section through a mountain by designating parts of the graph in Fig. 2.4 as the 'endemic ridge', 'the natural enemy ravine' and the 'epidemic ridge'. Southwood has used field data from a large variety of insects to plot such graphs of the way numbers between generations change in relation to density. He then arranged these plots on a third axis (stability of habitat) to produce a three-dimensional model (Fig. 2.5). This is rather like using the shape of the crusts of mixed-up individual slices of a sliced loaf of bread to reconstruct the shape of the loaf before slicing. Southwood has argued that the really pioneer insects seeking ephemeral niches outside the crop tend always to occur in the epidemic situation with crops, and that suppression (with insecticides) is the control technique. These are the *r*-strategists, insects for which the evolutionary pressure has been for increased rates of reproduction. However, if population growth of an *r*-strategist is slowed down by resource limitation (e.g. host plant resistance) the whole graph will move to the left and the natural enemy ravine can be deepened to hold the pest on the endemic rather than the epidemic ridge. At the other end of the diagram are insects which are unlikely to reach pest status because of numbers, but can cause economic damage at low densities (e.g. fruit pests and plant disease vectors, tsetse flies). These are the *k*-strategists, which have been selected for the ability to regulate their populations by adjusting mortality rather than increase rates. Southwood argues that the strategy against these is to disturb the habitat to a lower level of stability, perhaps possible by forms of cultural control such as soil cultivation or the removal of some diversity. An example might be the former practice of removing vegetation to make habitats unfavourable for tsetse fly survival. Also, stiletto control methods such as trapping should be effective against *k*-strategists, as indeed they have proved to be against tsetse flies.

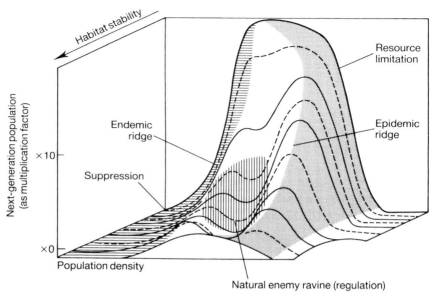

Fig. 2.5. Synoptic model of population growth/density relationships with pest management strategies superimposed (Modified from Southwood (1975); courtesy of Academic Press).

It is the insects from habitats of intermediate stability that are most affected by the 'natural enemy ravine'. Such insects are normally regulated by their natural enemies in the endemic situation. However, any sudden change (especially in density-independent factors such as climate and host plant nutrition) enables density to rise sharply. This enables the population to escape that regulatory restraint and then it will build up irrevocably on the epidemic ridge. This continues until eventually the pest population's own responses to crowding slow the population increase at very high densities. On the basis of this model, we can now seek to explain why outbreaks of crop pests and disease vectors occur.

Epidemic situations

Although we have just used 'epidemic' in contrast to 'endemic' above in a general ecological context, the word is widely used in the medical context to describe the situation where there is an unusually large number of people infected with a disease. Another term, 'epizootic' is used for a similar situation with diseases affecting animals or for massive outbreaks of crop pests. Essentially, the two terms mean the same as each other.

Increase in food/host resource

Monoculture with intensive inputs is a major cause of crop pests being elevated from endemic to epidemic status. Although these days it is hard to find any natural vegetation in Britain of the kind in which insects evolved, inspection of our roadside verges, hedges, commons and woodlands makes it clear that insect herbivores exist there under conditions where they do not occur in vast numbers, and the plants do not appear to suffer extensively from contact with them. We can contrast this 'endemic' situation with the 'epidemic' situation we often find in our crops.

Outside the crop, there will be a high mortality with insects failing to find their host plants in the mosaic of vegetation. Those that succeed will find plants variable in age and palatability, in soil providing poor nutrition and with strong competition for nutrients, light and water from adjacent plants. The plants will also have well-expressed chemical defences against attack. The potential population increase rate set by the density-independent factors will therefore be low. Prey colonies found by natural enemies are therefore not likely to escape their impact, and extinction on the plant is the probable outcome. With the poor nutrient status of wild plants, resources will be depleted early, resulting in a low threshold for overcrowding; emigrating individuals will have a low probability of finding a new suitable host plant.

Volatiles from the large area of a monoculture, on the other hand, make it easy to locate from a distance and arriving pests find that the density-independent factors allow a far more rapid growth rate. The plants have been bred, sown and managed to be uniformly suitable, and fertilizers and irrigation maintain this suitability for the whole season. Many of the chemical defences of the wild ancestors of the crop will have been reduced or removed during the breeding programme they have undergone, since such defences are at worst toxic to man and at best unpalatable. Harvesting of the previous crop usually means that density-dependent restraints in the form of natural enemies have to locate the crop anew and come in from outside. Most will only do this when prey outside the crop has become scarce and the crop is attractive, with the production of volatiles associated with the presence of the pest. They may therefore arrive too late to prevent inverse density-dependence, even if they are not killed with insecticides. These in turn depress pest populations to levels close to undercrowding, ensuring that their rate of increase is kept at maximum. Thus it is not surprising that pest populations are moved by monoculture from an endemic to an epidemic situation.

Movement of man, crops and insects to new areas

This can often cause outbreaks of pests as a result of changes in both the density-independent and the density-dependent restraints.

Panzootics of rinderpest, a viral infection, swept over much of Africa, especially southern Africa, during the latter years of the nineteenth century decimating the numbers of wild animals. Tsetse flies, such as *Glossina morsitans*, fed on these animals and as a consequence the fly's population crashed. After rinderpest had receded, wildlife gradually returned as did the tsetse fly. By 1913 there were many foci of tsetse flies, and by 1929 recolonization was at the rate of about 2600 km^2 a year, but some areas were not re-invaded until the 1960s when their advance threatened livestock production in areas that had been free of flies since 1895.

Malaria epidemics usually occur in areas where malaria transmission is unstable, that is where the level of transmission is periodical rather than static. After the invasion of Brazil in the 1930s by *Anopheles gambiae* from Africa there followed in 1938 an epidemic of about 200 000 cases and at least 14 000 deaths. This same mosquito species invaded lower Egypt in 1942 and caused about 160 000 infections and more than 12 000 deaths.

A herbivore at a low endemic level in a new continent may find the introduced crop allows a high population increase rate and that its local natural enemies only then show inverse density-dependence or are extinguished by insecticides used to control the pest. A good example is that of Colorado beetle, *Leptinotarsa decemlineata* (Fig. 2.6), and the introduction of potatoes to the New World as a crop monoculture by European settlers (see Chapter 1).

Resettlement of humans, either temporarily or permanently, has always happened and still continues, probably to an ever-increasing extent. Indonesia consists of more than 13 000 islands, Java being the largest and most populated. To prevent its further overpopulation there has been transmigration to other islands, sometimes coerced and euphemistically called colonization: the earliest records go back to 1905. Because of effective control measures there is little malaria in Java but it flourishes on many of the islands and has infected the transmigrants, some of whom have returned to Java and brought the disease with them. Rather similarly filariasis is not a problem in Java, or Bali, but it is on many of the islands on which farmers have settled.

Agricultural practices and also gold and gem mining activities, involving human migration, colonization and settlements on the fringes of forests in

Fig. 2.6. Adult (above) and larvae (below) of the Colorado beetle (*Leptinotarsa decemlineata*) (Courtesy of R. Coutin, © INRA, Paris).

Latin America, are one of the major causes of increased malaria transmission. For example, in parts of the Amazon region, during their first years tenant farmers start clearing and cultivating land that is inaccessible in the rainy season. They live in poorly built houses, or just in sheds, and are consequently very exposed to biting from anopheline mosquitoes, which in the rains breed in great numbers. Not surprisingly the farmers get malaria, often known as 'frontier malaria'. In areas of rice cultivation in Peru there has been increased malaria transmission. A problem here is that immigrant workers contract malaria when working on the farms and then return with it to the highland areas where it spreads amongst the community. Similarly in Venezuela, high malaria transmission on the Colombian border is partly due to immigrant workers coming from Colombia to work on the farms, while in Belize malaria is linked to seasonal migration of workers from El Salvador, Honduras and Guatemala to pick bananas and citrus fruits.

In various parts of the world people are rehoused on resource development projects, such as irrigation schemes. Often they do not own their houses and as a consequence these tend to fall into disrepair; communal latrines become broken, garbage and polluted waters accumulate and settlements turn into shanty towns. Under these conditions there are often hordes of flies which aid transmission of diarrhoeal infections, and water accumulates in latrines and cess pits, all of which provide ideal breeding places for *Culex quinquefasciatus*, a mosquito vector of filariasis. Settlers may be infected with malaria, filarial worms or arthropod-transmitted viruses and thus bring these diseases into an area. On rice irrigation schemes there will be plenty of mosquitoes to spread these diseases to the entire community. On the other hand immigrants may not have been previously exposed to diseases such as malaria and therefore have no protective immunity; consequently when they become infected they become very ill.

Movement of people for other reasons has often spread vector-borne human diseases to new areas. Louse-borne typhus epidemics have had a profound effect on many communities. Between 1917 and 1923 it is estimated that some 30 million people contracted typhus in eastern Europe and 3 million died. These epidemics were probably started by soldiers infested with body lice (*Pediculus humanus*) returning home during and after the First World War. During the Second World War (1939–45) Naples was in ruins and poverty, slum and unsanitary conditions were rife, just the conditions that favoured the proliferation and spread of body lice. When the Allied forces landed in Italy in 1943 there was a typhus epidemic and the death rate was as high as 81%; typhus threatened to virtually wipe out the city's million people. The epidemic was halted by delousing people using the then new insecticide DDT.

The arboviral disease dengue infects people and is spread mainly by the mosquito *Aedes aegypti*. It was originally an Asian disease, but after the Second World War the spread of *A. aegypti*, often through the species hitching lifts on ships and aeroplanes, and increased jetting of people around the world, including infected individuals, introduced dengue and the more lethal form, dengue haemorrhagic fever (DHF) into new areas. An example is the spread of DHF into the Americas starting with Cuba in 1981 when there were 344 203 cases of dengue, of which 1100 were apparently DHF. In 1988 there was a large epidemic of dengue in Ecuador involving an estimated 420 000 people.

More often perhaps, an insect regulated on the endemic ridge in its country of origin may cause epidemics in a new environment. This may be due to its having left its density-dependent restraints behind, and restoring these by introduction of natural enemies from abroad is the aim of classical biological control (see Chapter 7). Even in the absence of such intervention, and perhaps often supplementing them to a greater extent than realized, local natural enemies usually take some years to appreciate and adapt to the new food resource that has arrived, but will usually eventually do so. No classical biological control was undertaken against the lupin aphid (*Macrosiphum albifrons*) when it appeared in the UK from North America, yet five years later parasitoid mummies appeared in the colonies as well as ladybirds and other indigenous predators.

However, relaxations in density-independent restraints are often also involved. A more favourable climate in the new country may allow the pest to breed faster, but also local crop varieties are likely to be especially susceptible to the new pest as they were bred in its absence.

The spotted alfalfa aphid (*Therioaphis trifolii*) mentioned in Chapter 1 was introduced into Mexico from Europe in 1954. It was separated from its natural enemies and found the American lucerne varieties provided a faster breeding rate, especially in the warmer subtropical climate. Parasitoids were imported from Europe as part of one of the first integrations of biological control with insecticides (see Chapter 13), but later the breeding of new lucerne varieties less susceptible to the aphid provided a major contribution to control. Similarly, when the blue-green alfalfa aphid (*Acyrthosiphon kondoi*) reached Australia in the 1970s, the local lucerne varieties, which had never before been exposed to the pest, collapsed under the new invader. However, indigenous natural enemies eventually started controlling the pest and it was no longer a problem by the time plant breeders had developed an aphid-resistant lucerne variety for Australia.

Changes in man's management practices

A good example of how crop pests in the endemic situation became epidemic as a result of changes in crop management is the change in the 1970s from predominantly spring sown to autumn sown cereals. In grassland there was a typical endemic insect, the grass and cereal fly (*Opomyza florum*), existing at low population levels ovipositing in late summer in competition with other stem borers for grass stems in a suitable condition. As soon as wheat seedlings became available in monocultures at that time of year, the insect was elevated to epidemic pest status.

Similarly barley yellow dwarf virus was not a problem in spring sown cereals, since after inoculation by spring migrating aphids there was not time for the virus to replicate to damaging levels. However, autumn sown cereals became colonized by the autumn migration of a different aphid vector (*Rhopalosiphum padi*) and, with the extra months to replicate in the wheat seedlings, the virus was elevated to the most serious pest and disease problem of cereals with the aphid multiplying to epidemic levels.

Man also has had a dramatic effect on tsetse fly populations. As people cut down forests and clear land for farming, the shaded environment favoured by tsetse-flies is destroyed and their populations are pushed back. Also, in densely populated rural areas, cattle can often be kept without fear of their becoming infected with animal trypanosomiasis (nagana).

Changes in river management and poor clearance of weed to which the immature stages of the black fly attach have been implicated as among the causes for outbreak years of the simuliid black fly (*Simulium posticatum*), popularly called by the British press the Blandford fly. This fly breeds in the River Stour that runs through the town of Blandford, Dorset, in southern England. In 1969 suddenly it became a pest, biting people in the town and causing them to seek medical treatment; however, in 1973 there was a protracted lull, but large outbreaks resumed in the mid 1980s causing hundreds of people to suffer allergic reactions to the bites of this black fly.

There have been resurgences of dengue fever in many areas of the world. For example, following an absence of over 35 years, dengue outbreaks occurred in China in 1978 and 1979–80, while in 1985–6 an epidemic of DHF occurred for the first time. Among the reasons for the spread and resurgence of dengue in many areas is the rapid growth of urbanization, often accompanied by an unreliable water supply which leads to the proliferation of domestic water containers (Fig. 2.7) in which the vector, *Aedes aegypti* breeds. Increased air travel and deterioration in vector control also favour the spread of dengue. In the 1940s and 1950s an *A. aegypti* eradication programme in Latin America freed 19 countries of the vector. But because the mosquito was not eradicated from all countries and because the control programme deteriorated in the 1970s, re-infestations occurred. In 1995 the distribution of *A. aegypti* in Latin America was similar to that of the 1940s, and because of its reintroduction dengue fever has been reported from 43 countries between 1975 and 1995. Figure 2.8 shows the resurgence of the vector.

The use of insecticides is a management practice of man which can itself lead to epidemic resurgences of problems thought to be under control. Such resurgences often result because insects show tolerance to insecticides (see Chapter 5) or because the insecticide has destroyed natural enemies.

Fig. 2.7. Water-storage pots in Nigeria, typical larval habitat of the dengue vector, *Aedes aegypti* (M.W. Service).

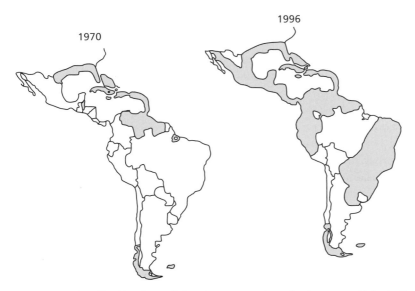

Fig. 2.8. The distribution of the dengue vector, *Aedes aegypti*, in Latin America in 1970 and 1996 (From Gubler and Kuno [eds.], 1997).

By killing the vectors, mainly *Anopheles culicifacies*, by spraying insides of houses where adults rested malaria eradication was being achieved in India. By the 1960s, just an estimated 0.1 million cases remained. In the 1970s, however, there was a resurgence of malaria. Some 7–10 million people were infected and the idea of eradication was abandoned. Similarly in Sri Lanka there were an estimated 2.8 million cases of malaria in 1946, but a DDT-spraying campaign reduced this to 17 detected cases in 1963. DDT-spraying ceased in 1964, because it was considered too costly when there was now virtually no malaria. Thereafter there was widespread resurgence and between 1968 and 1970 some 1.5 million people were infected. Renewed control reduced this figure in 1971, but in 1975 there was yet another resurgence. What were the reasons for such malaria resurgences in India and Sri Lanka? DDT resistance was partly to blame but it seems that a variety of socio-economic reasons, such as apathy, poor surveillance for vectors and the disease, and reduced budgets, were the main cause.

In certain towns in Sri Lanka, and in some other countries, insistence by health authorities that pit latrines are dug to combat diarrhoeal infections has often been accompanied by massive upsurges of *Culex quinquefasciatus*, due to its breeding in flooded pit latrines. This mosquito is an urban vector of bancroftian filariasis, and as a consequence there were resurgences of the disease.

A classic case of resurgence of a crop pest occurred in the 1950s, when the new organophosphate insecticide para-oxon was tested on many farms in the south of England. Natural enemies were wiped out, and re-invading cabbage aphids (*Brevicoryne brassicae*) then found fields free of this restraint on their multiplication, resulting in 'the most enormous cabbage aphid outbreak . . . ever . . . seen in England' (Ripper, 1956).

Climate changes

Climate, as a principal density-independent factor in insect population dynamics, has a major effect on the potential increase rate of both pests and natural enemies. In 1958 exceptionally high rainfall in Ethiopia resulted in explosive outbreaks of a sibling species (*Anopheles arabiensis*) of the *A. gambiae* complex; as a result more than 3 million people became infected with malaria, of whom 150 000 died.

The distribution range of host-specific natural enemies is inevitably smaller than that of the prey, so that any extension of the area at its climatic extremes will initially be free of natural enemy activity, and this will often lead to pest or vector outbreaks.

The debate continues as to whether there are real climate changes and what has caused them, or whether the perceived changes in some areas, such as increased rainfall and floods or higher temperatures, are just temporary cyclical changes. Nevertheless there is documentation of changes in pest distribution and disease transmission that have apparently been caused by climate changes, albeit possibly just short-term ones, and likely scenarios can be presented for long-term climate changes. If there is climate change during the present century it is bound to change the distribution of many pest and vector species, and a great deal of research currently focuses on modelling the possible consequences on pests of different climate change scenarios. Clearly pests will reach areas in which at present they never reach epidemic numbers just as they will probably become less important in others. The trouble is that as yet it cannot be predicted which climate change scenario will pertain where and when! Nevertheless it has been estimated that average global temperatures will have increased by 1.0–3.5 °C by the year 2100. The greatest effects on pest distribution and disease transmission are likely to occur at the extremes of the range at which arthropods can survive and disease transmission occur. For example, with vectors of human and animal infections the critical temperature ranges will be about 14–18 °C and 35–40 °C. A slight increase in temperature at the lower range will have a significant non-linear impact on increasing transmission, while at the higher end transmission could cease. But below this if temperatures rise to about 30–32 °C many pests will have faster life-cycles, and so more generations a year, and development time of pathogens and parasites in vectors will decrease. With tsetse flies an increase of 2 °C would likely cause the disappearance of flies from Central Africa, but an increase in numbers in other forested areas. Long-term changes in rainfall can also affect population dynamics of pests, and consequently disease transmission. Increased rainfall may provide more larval habitats for mosquitoes, especially in semi-arid areas, leading to the production of more pest species and vectors. Increased precipitation may also lead to denser vegetation and thus more resting sites for vectors, such as tsetse flies. Populations of disease reservoir hosts, such as rodents, may increase due to greater production of food crops, and this can favour transmission of diseases, such as plague. Human demography also plays a role in disease patterns. For instance climatic changes may lead to increased urbanization and more vectors such as the mosquitoes *Aedes aegypti* and *Culex quinquefasciatus*; conversely if marginal lands are occupied and farmed, this may extend the range of rural vectors such as many anopheline species of malaria to former cooler areas. For example, there has recently been malaria transmission in the Kenyan highlands, which are normally free of the disease. In the UK there are anopheline mosquitoes that are capable of transmitting some forms of malaria (remember that prior to the twentieth century malaria was common in many parts of the UK). Temperature is the main barrier to transmission. This is because at low temperatures mosquitoes must live longer for the life-cycle of the malarial parasite in the vector to be completed and this reduces the likelihood that they will live to that dangerous age.

During the 1997–8 El Niño phenomenon, a rise in temperatures and rainfall has been credited as being responsible for epidemics in Kenya of the mosquito-borne infections malaria and Rift valley fever. Since 1988 there has been an increase in monthly temperature of about 2 °C in the highland areas of Kenya, historically virtually free of malaria, and this temperature has been accompanied by several malaria epidemics. Other apparently climate-related epidemics have been observed in Tanzania and Rwanda, although some scientists still feel there is little good evidence that global warming has caused disease epidemics. Although rainfall has increased in some areas of Africa, mainly East Africa, reduced precipitation has occurred in parts of West Africa. For example, drier conditions in Senegal have led to the virtual disappearance of the vector *Anopheles funestus*, and over the past 30 years about 60% reduction in malaria transmission.

It can be notoriously difficult to predict the outcome of climate changes on pest abundance and disease transmission. For example, during the El Niño–Southern Oscillation in northern Brazil there were dry conditions and this brought reduced malaria transmission, although afterwards normal transmission levels were resumed. In contrast in Bolivia, Ecuador and Peru heavy rains accompanied the 1982–3 El Niño, as they did in Paraguay to Argentina during the 1991–2 El Niño, and in both instances there was increased malaria transmission.

None the less, most models predict that global warming would increase agricultural pest problems, and from these pest models an increased grain loss of 10–15% has been predicted for the more tropical countries of Africa.

The most convincing effects of climate changes on disease transmission are seen in Asia. Increases in temperature, rainfall and humidity in some parts of northern Pakistan have been linked with an increase in malaria transmission. In northeast Punjab, malaria epidemics increased five-fold in the year following an El Niño event, while the risk of malaria epidemics in Sri Lanka increased four-fold during an El Niño year.

As pointed out earlier, predicting what effect climate changes will have on pest numbers and disease transmission is made difficult by the many variables and processes linked to such changes, such as changes in human demography and land usage, and differences in resources to combat pest outbreaks. In essence the science of climate change and health is in its infancy. There is no shortage of research on effects on insects of climate change; the problem is that its value will remain a hidden asset until what climate change scenario is applicable where it can be predicted. Nevertheless, it seems reasonable to believe that if there are long-term climate changes, such as rise in temperatures and/or increased precipitation, then pest abundance and disease transmission will likely increase and/or extend to new areas.

3

Insecticides and their formulation

Introduction

The history of insecticide usage dates back many centuries, certainly to before 1000 BC, when the burning of sulphur to fumigate houses against pests and illnesses was mentioned by Homer. The insecticidal properties (pyrethrum, see later) of the flower heads of *Chrysanthemum cinerariifolium* (Fig. 3.1) were first known to the Persians, but were recognized in Europe only in the early nineteenth century. Another group of early insecticides were oils. However, the real landmark in terms of modern agriculture is the spread of the Colorado beetle (*Leptinotarsa decemlineata*) across the USA in the second half of the nineteenth century (see Chapter 1). Food production and the national economy were both threatened by this potato pest and, after much argument, it was finally decided to take the unprecedented step of spraying the potato crops in North America with a human poison (arsenic in the form of Paris Green). The mass human mortality predicted by the prophets of gloom did not occur, and there is no doubt that control of the Colorado beetle with Paris Green opened the way to a widespread use of biocides (destroyers of life in general) on crops destined for human consumption. In 1921 Paris Green was formulated as a dust that floated on the water surface and acting as a stomach poison killed mosquito larvae, mainly surface-feeding anophelines.

The use of Paris Green was followed by the development as insecticides of a range of existing compounds known to be toxic to man. Among these were plant poisons used by tropical tribesmen for tipping their hunting arrows, and eventually even synthetic insecticides, such as the organochlorines and organophosphates, based on chemical warfare research in the two World

Fig. 3.1. Pyrethrum (*Chrysanthemum cinerariifolium*) flowers (Ulead Systems).

Wars. As soon as the first synthetic insecticides became available, their cost effectiveness caused enthusiastic take-up of the new technology. The resulting overuse revealed, well within the first ten years, the major shortcomings of total reliance on synthetic insecticides – danger to humans and wildlife, persistence in the environment, resurgence of pest problems following destruction of natural enemies and the emergence of strains of the pest resistant to the toxins. However, the major alternative control methods have, for reasons which will be explored in this book, failed to convince farmers and public health workers to give up insecticides which remain, in spite of all the distaste the public have for them, man's principal weapon against his arthropod enemies. It is only fair to point out, however, that commercial failures never cause problems and that the problems of insecticides reflect their commercial success. Moreover, industry has responded to the problems by seeking to identify them before marketing, and many of the problems we associate with insecticides are far less likely with the insecticides of the last 30 years.

The industrial development of new insecticides

The search for molecules with new biological activity is an ongoing process in what are now relatively few multinational companies. With each year it naturally gets harder to find something new, especially something new clearly superior to compounds already on the market. Yet still each company typically collects about 100 new chemicals to test each month. These chemicals come from a variety of sources, including the company's own synthesis chemists, chemicals produced in other activities of the company (e.g. pharmaceuticals, paints, detergents), catalogues of suppliers of chemicals, and university Ph.D. theses. Increasingly the companies are buying into libraries of compounds created by combinatorial chemistry, where one material is chemically bound onto polymer beads, and reacts separately with other reagents. The process is then repeated with each of the resulting chemicals and a third reagent.

The identity and properties of some of the candidate materials obtained by this variety of routes may be unknown when they are put through the first stage of testing, the preliminary screen. Therefore very stringent precautions are taken for staff safety and for dealing with volatile and liquid effluents. The preliminary screen is a negative screen. This means that its aim is to reject the majority of compounds with zero activity against the biological targets tested, rather than to select some compounds as particularly promising. Although many of the chemicals will be tested simultaneously also for activity against fungal diseases and weeds as well as for any novel uses (e.g. accelerating plant growth), we will concentrate here on the insecticide screen. The candidate compounds are formulated sufficiently to permit spray application. Other forms of application are tested later during secondary screening. They are then sprayed at one high concentration against a range of pests representing the major feeding types of agricultural pests and selected medical/veterinary pests. The range will usually include a caterpillar, aphids, a beetle, spider mites, house flies and mosquito larvae. Increasingly the preliminary screen is moving to higher throughput techniques such as 96-well microtitre plates requiring only a few milligrams of the test compounds. Dispensing into these wells, and even assessment of the results, can be automated.

Any compounds showing insecticidal activity, perhaps some eight per month, are passed to the chemists for chemical processing. If necessary, the chemical is identified before a check is made that the company has not previously evaluated it. The compounds will be formulated (see later) for more effective application, and any known active parts of the molecule may be engineered for better toxicity (e.g. better penetration of the cuticle).

The formulated chemicals are then returned to the entomology section for secondary screening. This involves a large number of tests, many of which are very expensive; therefore results are regularly appraised to see whether continuing expense on further tests is really justified. The company may save a great deal of money if an early decision is taken not to proceed further with a particular candidate product. Some of the more important tests in the secondary screen are:

- Has the chemical an obvious market where it is better for its price than existing products? The cotton crop and mosquito control use most insecticide worldwide; it is often said that a chemical is only worth marketing if it has a use for one of these two purposes.
- Is the chemical safe on plants? Many chemicals are phytotoxic as well as toxic to insects. The chemicals are therefore tested at a relatively high dose on a variety of crop types; cucurbits are usually included, as they are particularly sensitive to toxins. Phytotoxicity is limited to visual symptoms a grower would notice; the checks do not include nutritional changes in the plant or mild checks to plant growth.
- How persistent is the compound and what are the breakdown products? Various parts of the molecule are radioactively labelled and the full pathway of breakdown monitored in plants, soil and water. Sometimes breakdown products can be as toxic as, or even more toxic than, the parent compound. So, as well as the parent compounds, the breakdown compounds will need to be tested for toxicity to humans and other non-target organisms (see below). The timing of residue decay is very important in relation to recommendations for the use of a chemical (see 'maximum residue level' below).
- Does the chemical pose a danger to humans? Clearly, if the insecticide is to be used against medical or veterinary pests then toxicity to humans and domestic animals is of paramount importance. These tests are by far the most expensive in the secondary screen. They aim to measure both acute and chronic toxicity to humans by extrapolation based largely on relative size (with an added safety factor) from the effects on mice and rats as well as on larger mammals such as dogs. Thus the oral and dermal LD 50s (the dose that would kill 50% of humans) have to be determined as well as any effects of long-term exposure of those applying the compound and of consumers to small quantities in food (chronic toxicity). The long-term tests also look for carcinogens and for mutagenic effects in mammalian reproduction. Since farmers, and occasionally public health workers, may apply more than one chemical at a time or in a short time interval, 'potentiation' must also be checked. The chemical is tested for human safety in mixture with a range of frequently applied products to check that the toxicity of each mixture is not greater than predicted from the toxicity of the components. From field tests following best application practice, the MRL (maximum

residue level) is calculated. This is regarded as the maximum permitted residue, and in agriculture determines the compulsory interval which must be left between applying a pesticide and harvest. Similarly there are strict rules as to how soon after livestock have been sprayed or dipped in insecticides they can be slaughtered for human consumption. A MRL must also be a safe level, and therefore must be below the maximum dose that would show no effect in humans divided by the safety factor of 100.

- Does the chemical have undesirable effects on non-target organisms? For an insecticide to have a sufficient market to be profitable, it has to be broad-spectrum. Thus DDT was a commercial dream because it is so effective against such a wide variety of agricultural, veterinary and medical pests and vectors. However, it is clearly a bonus if generally abundant natural enemies such as ladybirds, ground beetles and aquatic predators are not too badly affected. Toxicity to bees is also checked, and bird populations on the company's farms, where new products are tested in the field, are carefully monitored. Pesticides are applied to aquatic habitats, such as rivers and ponds, to control, for example, simuliid black flies and mosquitoes; agricultural pesticides may enter aquatic habitats through ground water or surface run-off. Adverse effects on fish and other aquatic organisms, including bioaccumulation in food chains (see Chapter 5) are therefore investigated. Also, soil studies are conducted to look at effects of the chemical on soil micro-organisms and the possible accumulation of pesticide along the food chain of soil arthropods, and similar studies may be conducted on plants, caterpillars and fish in laboratory systems.

Very few chemicals (perhaps two every five years) survive secondary screening, and for these the company will then seek limited trials clearance from national registration authorities to enable farm-scale trials to be carried out. Independent clearance will have to be obtained in every country where the company may wish to register the chemical for use. Data from these trials will be needed to convince a national registration authority that the new insecticide is effective and safe under their particular environmental conditions. In agriculture the trials are very expensive, as the crop may not be sold but has to be destroyed, and frequently the company will have to rent land from local farmers for the purpose.

In 1960, the World Health Organization established the WHO Pesticide Evaluation Scheme (WHOPES) to evaluate pesticides for use in public health. In 1987 the procedures were updated, and now comprise four phases. These are in essence the phases already described above, developed by industry to satisfy government registration authorities in respect of any biologically active compound. Thus Phase I of WHOPES comprises laboratory studies, which include efficacy, persistence and toxicity to humans and non-target organisms. Phase II is small-scale field trials, involving application methods, persistence, safety considerations etc. As for agrochemicals, Phase III is medium- and/or large-scale field trials which include epidemiological studies, formulations, safety, ease of application and cost-effectiveness. WHOPES has adopted the *FAO Code of Conduct for Pesticide Distribution and Use* designed for agro-chemicals and has established the Global Collaboration for Development of Pesticides for Public Health (GCDPP) to promote the search for safe pesticides and application methods.

Finally, the company will feel it has all the data needed to submit a portfolio for registration for a particular use in a particular country. The registration authority will select certain tests and have them repeated by independent laboratories, and then decide whether to allow the product to be marketed. Then there follows close collaboration between the registration authority and the manufacturer, when interim specifications are drawn up regarding the technical product and safety recommendations (this is designated Phase IV of WHOPES).

The whole process of industrial development of new insecticides is hugely expensive and at time of writing, is probably in the region of US$ 120 million (£85 million) per new product. A typical cash flow for the process is given in Fig. 3.2. A new chemical must be protected by patent early in development. The patent will last 20 years, so the company must have made its profit within 16 years of discovering the chemical, because after that date the chemical can be synthesized and sold more cheaply by other companies which have not had to bear any of the costs of development. Since the concept of 'profit' has to take into account the bank interest the company has lost by spending its money on insecticide development instead of investing it in savings, Fig. 3.2 makes it clear that a company must seek to take a new product to market after only 6–7 years of development. Since this is rather a short time to be certain there are no effects over the longer term, the company will continue to monitor the product on its farms after marketing, and may therefore need to withdraw the product after some years of sales. This should not alarm us; the company has 6–7 years lead time over usage by farmers or public health workers, and has identified a potential problem in good time. So when we hear of such a withdrawal of a recently marketed compound, we should applaud the evidence that the system works!

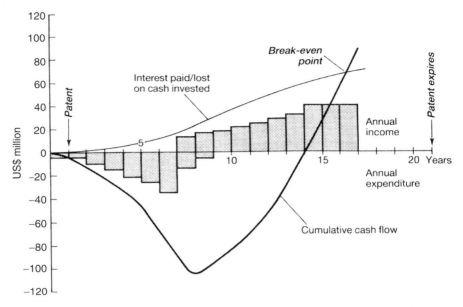

Fig. 3.2. Cash flow in the development of a hypothetical new insecticide, with a break-even point less than five years before the expiry of the patent.

The main groups of insecticides

Until recently, it was possible to include nearly all insecticides under a few groupings, but random testing by industry has now identified insecticidal activity in a large number of compounds outside major chemical groupings. None the less, the major groupings still represent the majority of insecticides in current or past use. Earlier insecticides such as DDT have been, or are being, withdrawn from use, particularly in the more developed countries. However, we will take a world view and select insecticides for mention to exemplify their variety and make points; mention of a chemical should not be taken to mean that its use is still generally recommended or permitted.

Before discussing individual insecticides, it is worth referring to the relevant nomenclature. The molecule in Fig. 3.3 is an insecticide recently banned in the UK, and any competent organic chemist could derive its chemical name as being 2'3-dihydroxy-2'dimethyl-7-benzofuranyl methylcarbamate. This is too much of a mouthful to be generally useful, so each pesticide has an internationally accepted common name, which for the molecule in Fig. 3.3 is 'carbofuran', a mainly agricultural compound. However, this will be in small print on the commercial formulation, since each of the several manufacturers of

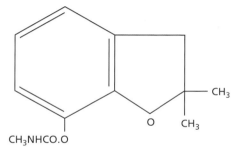

CH₃NHCO.O

Fig. 3.3. Chemical structure of carbofuran.

carbofuran will wish uniquely to identify their own product with a trade name. For example, carbofuran from the FMC Corporation is called Furadan, Bayer call their carbofuran Curaterr and Sanonda call theirs Afrofuran. Similarly, fenitrothion used for agricultural, veterinary and medical pests is variously called by its several makers Cekutrotion, Dicofen, Farmathion, Fenitron, Folithion, Nuvanol, Shaminliulin and Sumithion.

We will use the internationally accepted common name throughout this book, but bear in mind that this is not the obvious name on the tin! Additionally, commonly used alternative names (the usual 'trade' names) are given at the back of this book (Appendix of some chemicals and microbials used as pesticides).

Many insecticides have the same target in the insect, the nervous system. In principle, most insecticides will kill most insects in laboratory bioassays. The variety of insecticides available, however, does reflect a variety of routes whereby the toxin reaches the insect.

Routes to the insect

(a) Contact – The insecticide makes contact with the exterior of the insect, and penetrates the cuticle to reach the internal tissues. Ephemeral contact poisons are so short-lived that they need to contact the insect at the time of application.

(b) Residual – These are longer-lived insecticides which form a residue on the surface to which they are applied, from which the insect can pick up a toxic dose, usually by contact (residual contact poisons). The length of time the residue remains effective can vary greatly with insecticide, dose and with environmental conditions, from a day or two to many weeks for crops, but sometimes up to six months for house-resting mosquitoes (because the UV light from the sun which is so important in degradation of insecticide residues is greatly reduced in houses – see later).

(c) Stomach poisons – These have to be ingested by the insect to be effective. Often the residue is non-toxic, and enzymes in the gut of the insect release the toxin. With crop pests a stomach poison has a major advantage over a contact poison that it is 'addressed' only to a pest consuming the sprayed substrate (usually a leaf or fruit), and predators can move safely over the residue. However, there may be less specificity when aquatic habitats are treated, such as with the stomach poison Paris Green to control mosquito larvae, because other invertebrates may ingest the insecticidal particles.

(d) Translaminar – Insecticides with this property will pass through the leaf at the point of application to form a residue on the lower as well as on the upper leaf surface. Given the fact that most spray will land on the upper surface, yet most pests feed on the lower surface of a leaf, translaminar action is a very valuable property.

(e) Systemic – The chemical is absorbed by the plant (often to the point that the sprayed surface is no longer toxic) and moved internally in the vascular system to other parts of the plant. With almost no exceptions, this movement takes place only in the xylem, i.e. upwards and outwards in the plant. Effective protection can therefore often be given by dosing the soil around the roots rather than spraying the leaves. In spite of the xylem movement, phloem feeders (especially aphids) imbibe the poison.

Although not so widely used as they are against crop pests, systemic pesticides are sometimes employed to control veterinary pests. The chemical enters the animal's blood so that blood-feeding insects are killed. Clearly such chemicals must not be toxic to the host animal, and if this is destined for human consumption then there must be no toxic residues remaining when it is killed.

(f) Quasi-systemic – This is a rare but most valuable property. The sprayed substrate absorbs the material at the place of application rather like blotting paper, but does not move it to other parts of the plant. A quasi-systemic may therefore reach concealed feeders like stem-boring or leaf-mining larvae.

(g) Fumigant – The insect is killed by inhaling toxic vapours through its spiracles. Fumigants therefore tend to be the most effective ovicides since eggs, though well protected from contact poisons by their eggshells, still need to breath through their micropyles. Fumigants will also reach insects concealed in structures with openings such as tunnels and rolled leaves and kill, or flush out, insects sheltering in inaccessible places, such as behind cupboards, kitchen stoves or machinery.

Early insecticides

These were mainly of three kinds, namely plant derivatives, petroleum oils and the heavy metal salts.

Plant derivatives

Among the earliest insecticides, as mentioned above, were toxic extracts of plants long used by primitive tribes to tip their hunting arrows or to bring fish to the surface of rivers and lakes. Best known of these substances are pyrethrum (from flower heads of *Chrysanthemum cinerariifolium*; Fig. 3.1), rotenone (a root extract of the derris plant) and nicotine (from the tobacco plant). Originally the flowers of *C. cinerariifolium* were grown in the former Yugoslavia and the product, known as Dalmatian Insect Powder, was used to control body lice (*Pediculus humanus*) during the Napoleonic Wars. In the mid 1850s the insecticide pyrethrum was extracted and concentrated from the flowers and used to kill a large variety of insect pests.

These plant extracts work in a variety of ways, poisoning either the nervous or respiratory system. They penetrate the cuticle of insects and are ephemeral contact insecticides which are very short-lived (hours). The insect therefore has to be contacted by drops of spray or, in the case of nicotine when burnt, a toxic smoke is inhaled by the insect. The short life of these compounds was initially seen as a disadvantage, but today this 'disadvantage' gives them a special role when crops need treating close to harvest. Many other plants contain toxic chemicals, and several are known to be very toxic to man, e.g. hemlock (*Conium maculatum*) and deadly nightshade (*Atropa belladonna*). So-called 'natural insecticides' derived from plants can therefore be every bit as deadly as chemicals synthesized by man. However, the word 'natural' is enough to endear plant-derived insecticides to many who are worried about using other insecticides, and the short life of these chemicals after spraying certainly imparts safety to the environment. There is considerable interest shared by industry in discovering new insecticides in plants; e.g. azadirachtin from seeds of the tropical neem tree (*Azadirachta indica*) has been researched as an insecticide useful for both agricultural and medically important pests since the early 1960s in many countries.

Oils

The second group of early insecticides were petroleum oils which kill insects and mites, and their eggs. They worked mainly by suffocation, but in addition some mortality stems from the toxicity of their hydrocarbons. Oil was one of the earliest insecticides used to control mosquitoes and malaria. Typically larval habitats were sprayed with kerosene (paraffin), diesel oil or other readily available oils. This practice, however, was largely superseded when specially formulated 'High-spreading oils' that contained surface-active agents (e.g. octoxinol) that increased spreading power became available. Then, in an effort to obtain even greater killing power, small amounts of residual insecticides such as DDT and later more ecologically friendly toxic chemicals (e.g. temephos) were incorporated. Most of these oils were phytotoxic and killed non-target organisms. Currently very few oils are considered ecologically safe to use as larvicides, being largely replaced with microbial, organophosphate, carbamate or synthetic pyrethroid insecticides.

Because of the phytotoxicity referred to above, oils were mainly used on dormant leafless plants such as apple trees over the winter. Oils are still used in desperation today when mites, particularly, show tolerance to other pesticides, and then they may even have to be used on leafy annuals in spite of the inevitable damage to the plant. Vegetable oils at low doses such as 5 ml/kg of seed have proved very cheap and effective for protecting stored legumes such as cowpeas from attack by bruchid beetles (*Callosobrochus* species) in developing countries. The oil suffocates any eggs laid on the seeds and also any adult bruchids among the seeds.

Monomolecular surface films

These are included here as an adjunct to oils, though they are not actually among the early compounds. They are non-ionic, biodegradable surface active chemicals such as lecithins and isostearyl alcohol which, when sprayed on water, create a monomolecular organic surface. This alters the surface tension and thus interferes with the respiration at the water surface of mosquito larvae and pupae, which consequently drown. Such monolayers are today sometimes used in preference to oils and insecticidal sprays.

Heavy metals

The third group (including Paris Green = copper aceto-arsenite) are the heavy metal salts. These are toxic radicals (e.g. of arsenic or fluorosilicate) formulated as salts of metals (e.g. lead or sodium). Such salts are relatively stable, and plants can be sprayed without damage from the poisons, which have very general biological activity by precipitating protein. As a stomach poison, the salt must be ingested by an insect before the free toxin (e.g. arsenic) is released in the gut following hydrolysis of the salt. However, the heavy metal salts are also rather persistent, and therefore there is risk of ingestion by man. Also, the metals on which the salts are based are undesirable long-term soil contaminants. Paris Green was a widely used insecticide in the early-mid twentieth century against important chewing pests such as Colorado beetle (see Chapter 1) and lead arsenate was the mainstay of control of codling moth (*Cydia pomonella*) of apples. Formulated as a fine dust that floats on the water surface, copper aceto-arsenite has been used for many years to kill surface-feeding anopheline larvae of malaria vectors. It has also been formulated as granules that sink to the bottom of ponds and kill bottom-feeding culicine larvae.

Residual contact insecticides (Fig. 3.4)

Organochlorines
At a time when the short-lived plant extracts were used as contact insecticides, there was considerable interest in the possibilities of longer-lasting crop protection with insecticides which had a long residual contact life, i.e. insects walking on the dried spray on the leaf would contact a lethal dose of pesticide. In fact, thinking at that time was that the longer the life of the residue, the better. Similarly, the idea that insecticidal deposits sprayed on the interior walls of houses would kill house-resting mosquitoes, bedbugs, triatomine bugs etc. and when sprayed on vegetation would kill tsetse flies for many months was regarded as a possible panacea for control of several vector-borne diseases. Thus there was enormous welcome for the first residual contact insecticides synthesized by chemists, the persistent organochlorines. The most famous (or infamous, depending on viewpoint) of these is DDT, first synthesized as a molecule in 1874 by the German scientist Othmar Ziedler. However, its insecticidal properties were not discovered until 1939 by the Swiss chemist Paul Müller, who was awarded the Nobel Prize for Medicine and Physiology in 1948. For the first time, man had a weapon against the malaria mosquito, the lice infesting him during long battle campaigns and cheap long-term protection for his crops. The chemical could not only be sprayed, but often could be applied much more simply as a dust, as a coating to the seed (seed dressing) or just harrowed into the soil. One of the first uses of DDT was in Naples in 1944 when Allied forces of the Second World War dusted people with DDT powder to kill body lice (*Pediculus humanus*) to control a raging typhus epidemic. Such application to humans was possible because DDT and many other organochlorines are not unduly toxic to man; indeed DDT is unusual in that there is no reliable report of a human death resulting from acute DDT poisoning, a record that hardly any other insecticide synthesized by man can claim.

The problem of organochlorines which has led to their phasing out is that they break down rather slowly. Some, like the soil insecticides aldrin and

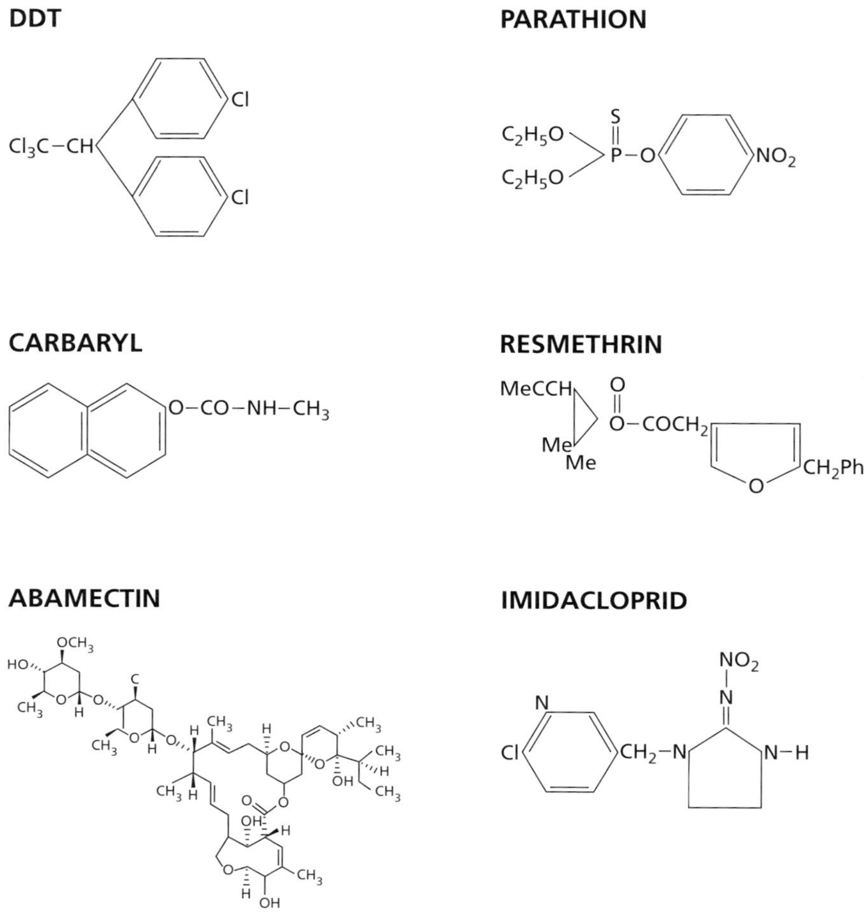

DDT

PARATHION

CARBARYL

RESMETHRIN

ABAMECTIN

IMIDACLOPRID

Fig. 3.4. Chemical structure of important groups of synthetic insecticides. Organochlorine (DDT), organophosphate (parathion), carbamate (carbaryl), synthetic pyrethroid (resmethrin), avermectin (abamectin) and nicotinoid (imidacloprid).

dieldrin, probably remain in the environment for periods approaching the lifespan of man. The organochlorines dominated the 1940s and 1950s, and many are still in use today, albeit increasingly on a restricted basis. DDT is still widely used in the tropics because of its low acute human toxicity. For example, a poor farmer tapping a DDT dust-filled nylon stocking towards himself over his crop rows, and spray-men in anti-malarial campaigns using their hands to mix DDT powder with water before spraying houses, would be ill-advised to switch to a more modern, and likely more toxic, insecticide. Moreover, for poor countries, DDT is relatively cheap. However, DDT and several other organochlorines have been banned in many countries because of their persistence in and damage to the environment. Yet, as pointed out above, DDT has a good human safety record, but despite this governments sometimes ban it even where there is no environmental hazard. Thus, where it is banned, it may not be used for spraying against disease vectors in houses, even though this creates virtually no environmental contamination. Some countries which have otherwise banned the compound (e.g. Ethiopia, Namibia, South Africa, Swaziland, Botswana and India) do still permit its use for house-spraying. If alternative insecticides have to be used, these are usually more toxic as well as more expensive. Moreover, if DDT is not used, house-spraying has to be repeated every 3–4 months instead of 6 months, which increases labour costs and logistical problems.

HCH (formerly called BHC) has fumigant properties as well as being a good soil insecticide and endosulfan, which has been used widely on crops and sometimes in aerial spraying against tsetse flies, is a very good general insecticide for the tropics. Here it has been used for many years and has really not created problems. It is biochemically safe for many biological control agents (see Chapter 13) and, unlike all other widely used insecticides, it has hardly suffered from tolerant or resistant pest strains.

The mode of action of the organochlorines is multiple and complex. The two most important actions are an inhibition of the enzyme cytochrome oxidase, which mediates gas exchange in the respiration of all animals which use blood as a gas carrier, and a destabilization of the nervous system.

Organophosphates

A second group of residual contact insecticides was produced in the late 1940s – the organophosphates, again as a result of chemical warfare research. The organophosphates are usually highly toxic to man but easily broken down and much less persistent than the organochlorines. Parathion and malathion were among the earliest organophosphates. The former is extremely toxic and has been sprayed from aircraft in Africa to kill *Quelea quelea* because these seed-eating birds can be serious crop pests. The high mammalian toxicity of parathion makes it unsuitable for use in habitations against disease vectors, but methyl parathion (less toxic than the earlier ethyl parathion) has uses in pest control on crops. The organophosphate group has been explored thoroughly to produce an arsenal of many diverse, flexible compounds. Many (including malathion) are strongly fumigant; others show translaminar action. Some, especially dimethoate, are quasisystemic. Quite a number (including metasystox and disulfoton) are systemic. Such systemic action can compensate for poor initial coverage with the pesticide and is especially effective against aphids and several other groups which then suck a poisoned sap; yet the plant surface may be quite safe for other insects, including parasites and predators, to walk over.

The organophosphates were increasingly preferred over the organochlorines during the 1960s, and are probably still the most widely used insecticide group today. Their great variety of modes of reaching the insect is their greatest value; no other group of insecticides offers the same flexibility. It has even been possible to replace the organochlorine soil insecticides such as aldrin and dieldrin with organophosphates by incorporating the latter (often diazinon) in a granule on or in the soil. The relatively non-persistent pesticide is continually replaced in the soil as the granule dissolves and it is even possible to surface-coat the granules, to delay the start of their breakdown. Such granules can be scattered over ground where, when flooding of the ground promotes hatching of eggs of mosquito species (e.g. *Aedes* species) that can withstand desiccation, the granules will then dissolve and release their pesticide, which kills the newly hatched larvae.

Like the organochlorines, the organophosphates have a very non-specific mode of action on animals, whether insects or man. They combine with the enzyme cholinesterase, and thus inhibit the hydrolysis of the acetylcholine produced at the nerve endings to carry nerve impulses across the synapses. In poisoned animals, therefore, acetylcholine accumulates at the synapses, giving constant nervous stimulation resulting in tetanic paralysis.

Organophosphates have now been in use long enough to have been 'put under the microscope' for potential hazards. Unfortunately, partly as the result of ill health of farmers using organophosphate sheep dips and the shadow of possible carcinogenesis, there have been some withdrawals of registration for this otherwise very useful group of insecticides, and it is likely that further severe reductions in their availability will continue.

Carbamates

A third group of residual contact insecticides, the carbamates (derivatives of carbamic acid) were introduced in 1956 with the compound carbaryl. The persistence of carbamates lies between that of the organochlorines and organophosphates, as does the toxicity of the first carbamates produced. Carbaryl has been widely used for the control of caterpillars and other surface plant feeders and against a variety of medical and veterinary pests. Methomyl has good contact action, but is also fumigant and slightly systemic; as well as on crops it has been used in toxic baits for house fly control. The early carbamates were followed by some highly systemic compounds, some of which (e.g. carbofuran and aldicarb) are very toxic to man. There is also a very unusual systemic and fumigant carbamate, pirimicarb, which has very low mammalian toxicity and a biochemical selectivity for aphids and most Diptera. When used against aphids, although ladybirds and parasitoids are not killed, dipterous predators such as midges (Cecidomyiidae: Itonididae) and hover flies (Syrphidae) suffer badly. The sensitivity of Diptera to pirimicarb has not been exploited for controlling pest flies, probably because of the low persistence of the compound. The action of carbamates, like that of the organophosphates, is on the nervous system by the accumulation of acetylcholine at the nerve synapses. Rather than inhibiting the enzyme cholinesterase, however, they act as competitors with the enzyme for the substrate's surface.

Carbamates are the most recent group to have selective bans imposed in some countries, especially for the highly toxic and persistent compounds such as aldicarb and carbofuran.

Synthetic pyrethroids

The next group of residual contact insecticides were the synthetic pyrethroids. It was long the goal of insecticide chemists to synthesize and modify the natural pyrethrum molecule and impart several additional desirable properties such as photostability to increase persistence. Although partial success came with the synthesis of allethrin as early as 1949, the real success came at Rothamsted Experimental Station in the UK in the early 1970s. The first synthetic pyrethroids, combining high toxicity to insects with low mammalian toxicity and greatly increased stability, were announced in 1973, and since then many new pyrethroids have been synthesized and marketed. Cypermethrin, for example, is 300 times more toxic than DDT to insects but only 60% as toxic to man. The toxicity of different pyrethroids is related to the proportions of four isomers, one of which is extremely toxic to insects. This single isomer was marketed as deltamethrin, a very effective insecticide at low dose, unfortunately therefore devastating to natural enemies. The mode of action of the pyrethroids appears to be a physicochemical process on the nerve membrane, sufficiently similar to the action of DDT that insect populations resistant to DDT can show some cross-resistance to pyrethroids.

The similarity to DDT does not end there. Like DDT, the synthetic pyrethroids are purely broad-spectrum residual contact poisons; no additional fumigant, translaminar or systemic action has been found. When applied, therefore, these highly potent insecticides are often very damaging to natural enemies. Like DDT, pyrethroids can be an irritant and repel insects. This can cause problems in limiting the contact between the target insect and the pesticide. However, repellency also has the advantage that, for example, mosquitoes that are not killed by contact with permethrin-treated mosquito nets are at least likely to be repelled, thus enhancing the protection afforded by such nets. Moreover, predators are often more mobile than their prey, and therefore are more likely to fly in response to the irritancy before they have accumulated a toxic dose. Synthetic pyrethroids therefore have some potential for selectivity in favour of natural enemies, if the latter can reach an untreated surface (see 'band spraying', Chapter 13). None the less, pyrethroids are very toxic to natural enemies, and pest resurgence resulting from destruction of natural enemies and development of resistance to pyrethroids in the pest have occurred sufficiently often to stimulate second thoughts on exactly how and when these valuable insecticides should be used.

Avermectins

The avermectins are a group of insecticidal and antihelminthic compounds originally produced by fermentation from the soil-inhabiting bacterium, *Streptomyces avermitilis*. The most commonly used is ivermectin, a veterinary drug that became available for human use in the late 1980s. It is increasingly being used in veterinary medicine to treat nematode infections and arthropod ectoparasites, such as lice, jigger fleas (*Tunga penetrans*), mange mites (e.g. *Psoroptes* species) and scabies mites (*Sarcoptes scabiei*) and ticks. Ivermectin can be administered through food; for example, food lots for cattle or deer can be treated with ivermectin to control ticks. However, in the USA the Food and Drug Administration (FDA) requires that such treated food fed to animals intended for human consumption is withdrawn in sufficient time for all ivermectin residues in the animal to have disappeared.

The main role of ivermectin in medicine is to kill microfilariae, the immature stages of filarial worms, such as those causing river blindness (onchocerciasis) in humans which is transmitted by simuliid black flies, and mosquito-borne bancroftian filariasis (responsible for elephantiasis). Ivermectin is now routinely given once or twice a year to people living in a very large area of Africa where onchocerciasis is endemic; the eventual aim is that within 12 years there will be a programme of self-sustained community-based ivermectin treatment in Africa. The great advantage of using ivermectin instead of more conventional drugs is that it has very few toxic side-effects.

The avermectins (especially abamectin) have also been used in crops, particularly against larger pests such as caterpillars in situations where extensive tolerance to other insecticides has developed.

Formamidines

Chemicals in this small group of insecticides act on the nervous system of arthropods by inhibiting monoamine oxidase, which is needed to prevent the accumulation of catecholamine neurotransmitters. Thus they have a different mode of action from the other synthetic neurotoxins, which target neurotransmission by acetylcholine. Therefore the formamidines can be used where tolerance has developed in the target insect to inhibition of acetylcholinesterase. Examples of such tolerant pests where chlordimeform and amitraz have proved valuable include livestock ticks and eggs and young caterpillars of some moth pests in agriculture.

Nicotinoids

Imidacloprid is a novel insecticide introduced in 1990, which again acts on the acetylcholine system, but by the different mechanism of blocking the postsynapse nicotinergic acetylcholine receptors. It is a systemic insecticide with good contact and stomach poison action. It has good persistence and can be applied to the soil as well as by spraying. It is effective against insects resistant to other compounds, and has been used especially against plant-sucking insects such as aphids, though it is also effective against many other pests. It has received such good take-up by farmers that there is a real danger that resistance to this compound may not be long delayed.

Insect growth regulators (IGRs)

These are chemicals which interfere in an adverse manner with the normal growth and development of insects (Fig. 3.5), usually because of their close relationship to an insect's natural internal hormones, or by acting as antagonists to the latter. The idea of using such products originated from an insect endocrinologist, Carroll Williams, who suggested that the hormones produced internally by insects to regulate their moulting and metamorphosis could be turned back on them as 'third-generation insecticides'. Original claims made for such third-generation insecticides were that they should be highly specific, and that resistance to them would be unlikely. However, resistance to hormones and 'hormone mimics' has occurred, presumably because insects already possess the compounds to maintain their own hormone titre internally.

IGRs are relatively non-specific among insects, but certainly they are likely to be less damaging to other types of organisms in the environment than insecticides. They have very low toxicity to mammals, birds, fish and adult insects, but when sprayed onto waters for mosquito control they unfortunately kill crustaceans and immature stages of various aquatic insects. They quickly break

Fig. 3.5. Pupal deformity following application to large cabbage white butterfly (*Pieris brassicae*) caterpillars of a chemical with structural similarities to insect moulting hormone (Courtesy of Syngenta).

down in the environment, unless applied as granules or microcapsules. An advantage is that many IGRs kill late in the insect's life cycle, thus minimizing any reduction in density-dependent mortality (e.g. predation or cannibalism, see Chapter 2), a problem that arises with some biological control measures. This advantage is illustrated when IGRs are used to control mosquitoes. The delayed mode of action means that early stage mosquito larvae remain available as food for fish, wildfowl and other predators. This is not only of benefit for biological control; it is also a conservation bonus.

Fleas are especially sensitive to IGRs, and the use of these compounds to control fleas on cats and dogs is both popular and effective. Preparations are either applied topically to animals, injected subcutaneously, impregnated into plastic collars, applied to premises where animals live, or are mixed with the animal's food.

IGRs are relatively costly compared with more conventional insecticides and are not as widely available. However, they can be useful in the face of insecticide resistance or when insecticides cannot be used because of their environmental damage (see methoprene below). Few new compounds have been developed since the arrival of methoprene in 1974.

Ecdysone.   This is the 'moulting hormone' which regulates the moulting process and the development of the new cuticle. Although giving promisingly high insect mortalities in the laboratory, ecdysones have run into problems of commercialization, particularly because of the high costs of synthesis.

Juvenile hormone analogues.   These function in the same way as the natural juvenile hormone in the regulation of metamorphosis, and may or may not be similar to the natural hormone in chemical structure. Two such analogues are methoprene and kinoprene. The effects of these compounds are usually seen during larval to pupal metamorphosis, and various degrees of incomplete metamorphosis become apparent. Larval–pupal mosaics may be produced, or strange deformations may appear on the pupal structure. Other uses of juvenile hormone analogues are in disrupting embryogenesis in the eggs and in preventing adult diapause. They have been extensively tested in public health and stored products because of their relative safety to human beings. Methoprene has also been used to control cat and dog fleas and mosquito larvae. It is approved by the World Health Organization for use in drinking water. Briquettes or granules containing methoprene can be used in potable waters, as well as other aquatic habitats, to give up to about four months' control of mosquitoes.

Unfortunately, resistance in insects appeared very quickly to the few JH analogues that have been marketed. For example, resistance to methoprene was recorded in several *Aedes* mosquito species in the USA soon after it was introduced. Other commonly used JH analogues include fenoxycarb against termites and more rarely against codling moths (*Cydia pomonella*) and pyriproxyfen, commonly used for flea control.

Anti-juvenile hormones.   A high titre of juvenile hormone in the insect maintains its larval characteristics, and if this effect can be counteracted in the early stages, then the larvae may metamorphose into miniature pupae or sterile adults. This is known as precocious development, and the name 'precocenes' has been given to a major group of anti-juvenile hormone compounds. None of the compounds developed so far has been sufficiently active for practical purposes, but there is still hope for the future. One problem with precocenes is that they could actually increase damage as, if applied too late in insect development, they prolong larval life rather than the formation of precocious pupae.

Chitin synthesis inhibitors.   These are the most extensively commercialized and used IGRs. The benzoylphenyl ureas, which interfere in chitin synthesis in insects, were first discovered about 1970. Since then a number have been marketed quite successfully, particularly against the Lepidoptera. These compounds interrupt the organism's moulting process. Although the new skin seems to be formed normally, it is the shedding process that is disrupted, and affected insects either die within their old cuticle or fail to emerge satisfactorily from it. Though effective, the compounds are still rather expensive compared with traditional pesticides. Nevertheless, chitin synthesis inhibitors are particularly useful where some selectivity of action is required (parasitism is usually not reduced by application) or where the pest has become resistant to insecticides. Examples of these chemicals are diflubenzuron, lufenuron and triflumuron. Diflubenzuron is commonly used to spray mosquito breeding places, but it should not be used in drinking water. The same compound, though expensive, has been used for caterpillar control in orchards and whitefly control in glasshouses where tolerance to other insecticides has appeared and/or where the use of biological control demands a more selective product.

Triazines.   This very small class contains cyromazine, which causes stiffness of the insect cuticle. Death results from reduced growth and integumental lesions. Cyromazine is most effective against Diptera. In the USA resistance to cyromazine had developed in house flies after it had been used in poultry feeds for many years.

Formulations

For an active ingredient to be useful in practice, it has to be formulated with additives of various kinds so that it can be diluted before application and then applied effectively. Common additives are emulsifiers, wetters, spreaders and stickers. Formulation can greatly enhance the convenience of application and the biological activity of an insecticide, as it affects the adhesion of spray to surfaces, penetration of insect cuticles, persistence of the insecticide residue, shelf-life in store, phytotoxicity and safety to humans. Good formulation chemists are worth their weight in gold in industry; it seems almost as much an intuitive art as an objective science. Formulation chemists face huge variation in the properties of the active ingredients they are expected to formulate. The chemicals may be solids or liquids, they may be stable or totally unstable to air and light, volatile or non-volatile and they may be water soluble, oil soluble or even insoluble in anything. 'Formulation' is used both as a noun and a verb – formulation is the process of producing formulations!

Formulations applied as liquids

Water-soluble powders
These are usually sold as ground or pelleted solids with some water-soluble inert filler and a small quantity of wetter. Water-soluble materials tend to hydrolyse and decompose if sold in a liquid formulation.

Water-miscible liquids
These can be used where the chemical is insoluble in water, but will dissolve at least as 150 g/litre in a solvent such as alcohol which is itself miscible with water. Such formulations need testing that they will remain stable in storage and not separate out, particularly at low temperatures.

Suspension concentrates (flowables)
The active ingredient is milled to particles of less than 5 μm diameter together with dispersing and anti-settling agents to make a thick 'syrup' which can be thinned and sprayed by dilution with water. The chemical can therefore be extremely insoluble in any solvent. In fact solubility no greater than 1000 ppm is actually necessary; any higher solubility, and dissolved material will come out of solution and grow on the existing particles (a process called 'Ostwald ripening'). One advantage of this formulation is that it can contain a very high percentage (80%) of the active ingredient compared with other formulations.

Water-dispersible (wettable) powders (WP on labels)
This is probably the commonest spray formulation. Wettable powders can reach 30% of active ingredient by volume, and the chemical is blended with 5–20 μm particles of an inert filler such as talc or clay, which can absorb the chemical (usually in liquid form) up to 80% of the filler's weight. Wetting agents are particularly important additives to prevent the particles floating. With constant agitation in the spray tank, the powder can be diluted with water and the formulation emerges from the nozzle as a particulate spray. However, although convenient, the particles do cause greater wear on the nozzles than other spray formulations.

Emulsifiable concentrates (EC on labels)

This is probably the second commonest formulation, since it is more usual for insecticides to be soluble in organic solvents than in water. Normally a 25–50% solution in a solvent is marketed; at least 10% solubility is needed to make the formulation economic to transport. Emulsifiers are added to ensure that a fine oil drop (1–2 μm) in water emulsion is produced when the formulation is diluted with water. A good emulsion appears as an opaque white fluid instantaneously and does not settle out for 24 hours. Invert emulsions (forming water drops in oil) are also available, but are used entirely for herbicides. Sometimes an emulsion is sold as 'preformed', where the concentrate has the oil already dispersed in a little water.

Emulsifiable concentrates are a convenient way of formulating many water-insoluble ingredients and they do not cause nozzle abrasion. However, the solvents which make up a large proportion of the marketed product are expensive, and also lead to these formulations being potentially more phytotoxic than most others.

Oil formulations

These are solutions of ingredients (usually not water-soluble) in a low-volatility heavy oil. This formulation overcomes the problems of evaporation of small drops, but because of the weight and cost of oil, they are only suitable for controlled droplet application (see Chapter 4), which requires low amounts of concentrate compared with other application methods.

Microcapsules

Here the insecticide is sprayed as small sticky-polymer coated droplets which adhere to the sprayed substrate, but remain harmless to the plant and beneficial insects since the poison is only released once an insect has ingested a capsule (stomach poison action). They thus increase the selectivity of any compound suitable for such formulation, but they are not surprisingly rather expensive. Self-forming capsule formulations are now available. These pass through a spray nozzle as liquid drops, but thereafter contact with the air immediately causes the 'shell' of the capsule to form. This latter approach causes us to classify microcapsules as formulations for liquid application. In aquatic habitats microcapsules can sometimes be formulated so that they are ingested by targeted pests, but rarely by non-target organisms (see Chapter 13).

Paints and lacquers

Various commercial lacquers and paints can be made to incorporate residual insecticides. These can be painted onto walls and other surfaces, especially in kitchens and restaurants where insecticidal spraying is inappropriate because of health considerations, and remain effective in killing cockroaches for several months. Slow-release formulations of insecticidal paints based on latex or polyvinyl acetate can be applied to house walls to control triatomine populations.

Formulations for dry application

Dusts

These are the oldest formulation. Dusts are sold ready for application, usually comprising around 92% filler, such as kaolin and chalk, with 8% adsorbed active ingredient. They are therefore expensive to transport, but often they require no application equipment (they can if necessary be broadcast by hand). Sometimes, however, they are applied by a simple bellows-type duster. Such dusters are used to apply powders under peoples' clothing to kill body lice (*Pediculus humanus*) or to apply dust specifically to the central shoot of cereal plants against stem borers. An advantage in the tropics is that dusts require no water for dilution. Deposition on plants can be poor with drift commensurately high, and one group of insecticides (the organophosphates) cannot be used because they react chemically with the filler.

Granules

These are an increasingly popular formulation, being relatively safe to handle and easily applied, particularly if the insecticide has systemic activity. Because of the low danger of drift, they usually consist of twice the proportion of insecticide as do dusts, and may even reach 25% of active ingredient. Normal granules can either have the insecticide impregnated on a pre-formed granule or, more expensively, mixed with filler first and then extruded as granules. This latter type can form slow-release granules, which mimic persistent insecticide with less persistent compounds by continually releasing new molecules into the soil or water as the granule dissolves. Thus the pesticide is replaced as it is broken down. It is even possible (with either kind of granule) to apply granules as 'time-bombs' by coating the granules with a polymer which will destruct in a known time to start releasing the active ingredient.

Granules are usually 0.25–1 mm in size, but occasionally larger granules (1–2 mm), often termed 'pellets', are used on aquatic habitats. At the other extreme, microgranules of only 100–200 µm are also made as a 'low-drift' dust.

Briquettes

Insecticidal wettable powders are mixed with plaster of Paris and sawdust or with cement and sand to form briquettes (10–300 g) which, when placed in water (e.g. water butts, septic tanks), slowly release the insecticide which kills aquatic insects such as mosquito larvae. Because of its low mammalian toxicity when temephos is incorporated into briquettes, these can be used in water-storage pots containing water destined for human consumption.

Impregnated plastics

Plastic collars as worn by cats and dogs, and ear tags in livestock can be impregnated with insecticides or IGRs for their slow release over several months to control ectoparasites. Similarly plastic strips impregnated with insecticides such as the fumigant organophosphate dichlorvos (now banned in some countries, including the UK and USA) are commonly hung in dairies, restaurants, houses, cupboards, greenhouses etc. to protect against a wide range of pests and vectors.

4

Application of insecticides

Introduction

The subject of pesticide application involves some really fascinating topics such as the fluid kinetics of droplet production and the engineering aspects of spray outlets (nozzles) and pressure sources (pumps). Much of this, however, lies outside the scope of this book, and readers are referred to Matthews (2000) for an excellent and well-illustrated account.

Formulation and the method of application can have almost greater influence on the efficiency and selectivity of kill than the choice of active ingredient. How these variables may be manipulated so that the pesticide application is less damaging to natural enemies is discussed in Chapter 13. It is a long, long way in biological terms from the emission of pesticide from a machine to achieving kill of a pest. The first problem is to get the right amount of chemical onto the target. There can be many targets. Plant surfaces are not only crops or competitive weeds; pesticides may also be sprayed on uncultivated land. In the past blanket spraying of vegetation, especially riverine vegetation, with residual insecticides to kill the tsetse fly vectors of both human trypanosomiasis (sleeping sickness) and animal trypanosomiasis (nagana) which can devastate the livestock industry in sub-Saharan Africa, has been commonly practised. Before the persistence of DDT in the environment was recognized (see Chapter 5), hedgerows adjacent to crops were sprayed to kill roosting cabbage and carrot root flies (*Delia radicum* and *Psila rosae* respectively). Spraying aquatic habitats with insecticides to kill mosquito larvae (to control mosquito-borne diseases or alleviate mosquito bites) is still widely practised throughout

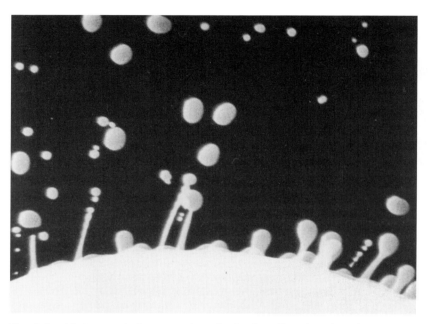

Fig. 4.1. High-speed photography of spinning disc, showing the fragmentation of the liquid sheet into filaments, which rupture to produce main and satellite drops (Courtesy of Syngenta).

the world – probably more effort and money is expended on this in North America than anywhere else.

Inert surfaces (particularly in buildings) are common targets in medical entomology. For example, residual house-spraying is undertaken to kill resting pests and vectors such as bedbugs, mosquitoes transmitting malaria, and in Latin America triatomine bugs transmitting Chagas disease. Indeed attempts to eradicate malaria (e.g. in India) have relied on this approach. Kitchens in hotels and restaurants may be sprayed to kill cockroaches, and the method can be adapted to kill flies and ticks sheltering in dairies, stables and barns. Domestic animals may also be targets sprayed directly (see p. 102).

Spray application to the target/surface

It is the weight (usually measured as size) of the drop, not when it leaves the machine but when it reaches the target, that is of critical importance. Unfortunately a spray cloud never consists of identically sized drops. Nearly

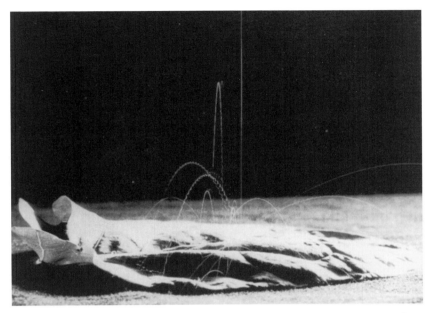

Fig. 4.2. Photographic track of droplets impinging on a reflective leaf surface and 'bouncing' off instead of adhering (Courtesy of G. D. Dodd).

all spray equipment used on the farm today, as it has always done, relies on forcing liquid through a hole under pressure to produce a spray, and this process (which amounts to disintegrating the edge of an expanding sheet of liquid) produces particularly variable drop sizes. High-speed photography (Fig. 4.1) shows that, at the edge of the sheet of liquid, larger (main) drops are produced on filaments which shatter to produce a larger number of small (satellite) drops. The larger drops contain most of the insecticide a farmer has paid for, but cover very little of the crop surface if they are retained on the foliage at all. It is quite normal for 60% of the total spray volume to be used for only 20% of the drops (the largest ones) formed at the nozzle. These large drops will often bounce off the leaf they contact, whether of a crop or vegetation harbouring tsetse flies, particularly if the leaf is hairy or very waxy. This 'spray reflection' may occur with drops about 250 µm in diameter or larger. This reflection, and the effect thereon of different leaf surfaces, can be simply demonstrated by passing leaves or leaf discs laid on filter paper under a source of coloured drops, e.g. a narrow-tipped burette containing a coloured solution, and leaving the tap very slightly open (Fig. 4.2). The other problem with larger drops that are actually retained on the leaf is that they are likely to spread into each other

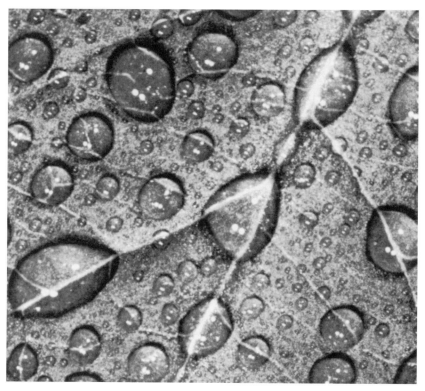

Fig. 4.3. Close-up of part of a sprayed leaf illustrating the distribution of different drop sizes in a spray. The smaller drops provide the bulk of the spray coverage (Courtesy of Syngenta).

and coalesce; when this happens, the liquid on the leaf collapses into a very thin film, and much runs off the leaf onto the ground.

It is therefore the smaller drops, vast in number but accounting for rather little of the total volume applied, which provide the coverage needed by the farmer on the foliage for pest control (Fig. 4.3). Table 4.1 shows how coverage (as drops per cm^2) decreases dramatically as drop diameter increases; this is of course because the volume of a drop is related to its diameter by the cube. However, the smaller drops are much less likely than the larger drops to come in contact with the foliage in the first place. Moreover, small drops have an enormous surface to volume ratio, and will rapidly evaporate and get even smaller as soon as they leave the nozzle, particularly in warm and dry weather. The problem with small drops is that, because of their small weight, they

Table 4.1. *Number of drops per cm² over 1 hectare of ground which could theoretically be obtained from a litre of liquid distributed as uniform drops of different sizes (from data of Ripper, 1955).*

Diameter (μ) of drops	Drops per cm²
10	c. 20 000
20	c. 2400
50	153
100	19
200	2.4
500	0.3

have very little momentum, and lose speed very rapidly. When we see a cone of spray apparently driving its way into the crop canopy, it is actually the large drops we are seeing; the small ones are mostly invisible to the naked eye. It is not too far from the truth to say that what we see coming out of a nozzle above a crop will mostly be wasted and finish on the soil. However, many of the invisible small drops will also be wasted, as they are too small. A good way of illustrating the problems of very small drops is to crush the end of a piece of chalk, and then to throw different sized fragments at the far wall of a room. It is quite easy to reach the wall with pieces the size of the letters printed on this page, but impossible with the really fine particles. This is because, like the small drops in a spray, they quickly lose any momentum we impart (due to friction with the air of a drop with a large surface to volume ratio), and come to a stop in relation to the air around them. If that air is stationary (it rarely is out of doors), gravity will take over, and the drop will fall under its own weight and 'sediment' onto whatever surface it first encounters. The smaller the drop, the more slowly it will of course fall, and it may well evaporate away to no more than a tiny particle of insecticide before it has made a contact. If the air is moving sideways, it will carry the stationary drop sideways during its fall. As the air approaches an obstacle, e.g. a plant stem or leaf, it will stream around the obstacle, carrying the small drop round as well. Even drops which still have a little momentum of their own will tend to follow the airstream, and may well then accumulate, if at all, on the edges of leaves or on leaves edge-on to the airflow. Such leaves may collect insecticide deposits seven times greater than leaves facing the spray. All this assumes that the small droplet has got to the vegetation in the first place. Often in daytime, however, air movement has an upward component resulting from the sun heating the earth; the warm air near the ground then rises to great heights (thermals). That is, of course, also the direction taken by very small drops under those conditions, and insecticide may then 'drift' very long distances, well away from the target area. Such drift will be most pronounced on a warm still day, when the smoke from chimneys rises vertically upwards. Because of thermals, aerial spraying is usually undertaken during cooler parts of the day, such as early mornings or evenings. However, spraying crosswind on a day with a mild breeze will increase the plane's swathe width and, whether from the ground or the air, spraying is best done with a crosswind than in still morning or evening conditions. Small drops sprayed with a mild crosswind are much more likely to finish in the crop or other vegetation being sprayed.

The droplet spectrum of a spray is therefore an important criterion. As the smallest drops cannot be caught on any targets, the only reliable measurements of the diameter of droplets in a spray cloud must be made while the spray is airborne, and this can be done with expensive laser diffraction equipment. From the diameters of a large number of drops, important calculations can be made. Imagine that the drops have solidified and magnified – in terms of relative size there will usually be a few footballs, rather more tennis balls, lots and lots of table-tennis balls and myriads of peas! Clearly there will be a diameter (probably somewhere in the tennis ball size range), where as much liquid volume is contained in a few larger drops as there is in the very many more smaller drops – this is the volume mean diameter (vmd). But there is also a second important diameter, the average drop size. More than half the drops may well be peas so this, the number mean diameter (nmd), will lie somewhere in that range. It will usually be much smaller than the vmd. Put crudely, the vmd is where the money goes and the nmd is what can provide the coverage of an area; the aim is not to have a vmd much larger than the nmd, and for crop spraying we seek an nmd of around 120 μm.

In the end, a very small segment of the droplet-size spectrum will both reach the target and be retained thereon. Thus drops in the 30–50 μm diameter range may be right for contacting insects resting on foliage, whereas 100–150 μm drops are more likely to deposit on and be retained by the foliage itself. Some idea of how small the appropriate segment is in relation to the cost of pesticide to the farmer can be obtained from the results of a test where a spray of Bordeaux mixture (a fungicide based on copper) was applied to cocoa leaves. The application, with 100% capture and retention of the spray, would have left a deposit of 25 μg of copper per mm² of leaf. The maximum deposit achieved was 1.3 μg/mm² on the lower surface of the young leaves; on the waxy surface of mature leaves the deposit fell to 0.7 μg/mm^2.

A typical amount of pesticide (quoted as amount of active ingredient, which is often a half or less of the volume of the commercial formulation the farmer purchases) applied per hectare is 750 g. The average deposit in the example just quoted would mean a farmer would be lucky to leave 30 g of the 750 g he has paid for on his crop plants; the rest would end up as a contaminant on the soil or would be carried away in the air. In practice, the farmer cannot usually match the very carefully controlled application used in the cocoa experiment. A more realistic figure would be that 10 rather than 30 g/ha would be deposited from a 750 g/ha spray.

Is this waste and consequent loss of insecticide to the environment inevitable? Unfortunately at the present time the answer is 'yes', although better maintenance of equipment and choice of conditions for spraying could no doubt cut the wastage to some extent. However, there have been developments in pesticide application technology (see 'spinning cage', p. 88) which hold the potential for future improvement of the situation.

Hydraulic sprayers

The statistics of wastage quoted above are typical of the most common form of insecticide application equipment, the hydraulic sprayer. Whether the machine is a knapsack sprayer carried on the back and pumped with a handle, a compression sprayer pumped up to pressure (Fig. 4.4) and slowly discharged or a tractor-mounted boom sprayer, the principles are the same (Fig. 4.5). Pumps of various designs for different pressures force liquid towards the nozzles, with a head of air compressed by the pressure of the liquid to even out the flow in spite of the pulses from the pump. The most common nozzles which form the sheet which is broken up into spray are flat fan and hollow cone nozzles. The orifice in flat fan nozzles (Fig. 4.6) is a narrow rectangular slit at the apex of a V-shaped insert, and the liquid fan-shaped sheet produced disintegrates into drops as the edge of the fan is stretched further away from the nozzle. Fan nozzles on a boom are normally offset by about 10° to prevent the edges of the fan colliding. Fan nozzles are very popular, they penetrate crops well and it is easy to get even coverage with several fans on a boom. One sees them used routinely to spray many crops and vegetation on which tsetse flies rest, also the walls and ceilings of houses and animal shelters to kill triatomine bugs, bedbugs and mosquitoes. However, although they produce a visually satisfying stream of liquid, unfortunately a lot of it is large wasteful

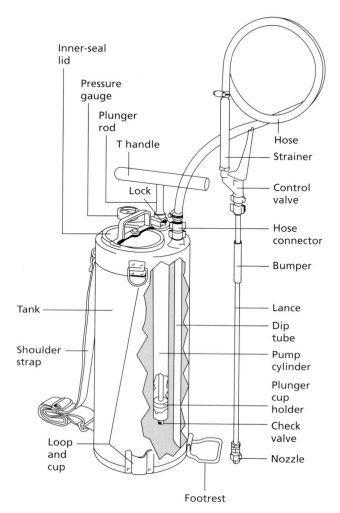

Inner-seal lid

Pressure gauge

Plunger rod

T handle

Lock

Hose

Strainer

Control valve

Hose connector

Bumper

Tank

Lance

Dip tube

Shoulder strap

Pump cylinder

Plunger cup holder

Check valve

Loop and cup

Nozzle

Footrest

Fig. 4.4. Cutaway diagram of an X-pert Hudson compression sprayer, the most commonly used sprayer used for spraying residual insecticides in houses to kill pests and vectors such as mosquitoes, cockroaches and triatomine bugs. Such sprayers are also used for spraying crops (Courtesy of H. D. Hudson Manufacturing Company, Chicago, Illinois).

drops. Indeed, the droplet spectrum of these nozzles is particularly poor; often the vmd is more than ten times the nmd.

With hollow cone nozzles it is much harder to get even distribution from a boom and the spray is finer. There are many small drops, but fewer large drops

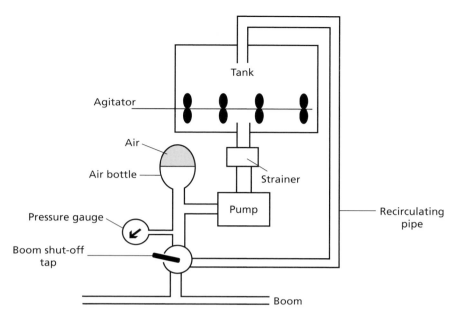

Fig. 4.5. Diagrammatic representation of principles of a tractor-mounted hydraulic sprayer; knapsack sprayers have most of the same features.

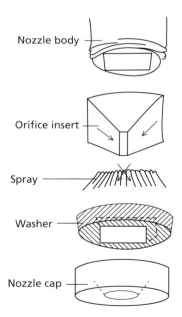

Fig. 4.6. Exploded structure of a flat fan nozzle.

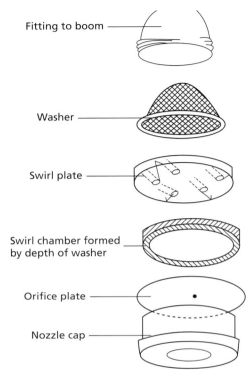

Fitting to boom

Washer

Swirl plate

Swirl chamber formed
by depth of washer

Orifice plate

Nozzle cap

Fig. 4.7. Exploded structure of a hollow cone nozzle.

than with fan nozzles. Vmd/nmd ratios are typically 4–6. The hollow cone is produced by the liquid having a vortex imparted by being forced through slots cut in an angle through what is called the swirl plate (Fig. 4.7) into the swirl chamber created by a washer between the swirl plate and the orifice chamber. Parts are interchangeable and different nmds and cone widths can be obtained by varying the size of the orifice, the depth of the washer and the number, size and angle of the holes drilled through the swirl plate.

Mistblowers

This type of machine, used for space-spraying houses as well as for spraying trees and shrubby vegetation, produces drops on the 'Venturi' scent spray principle (Fig. 4.8). A motorized fan creates a powerful air-blast in a length of trunking into which liquid is dribbled under gravity. The air-blast produces a sheet of liquid from the end of the dribble-tube, the edge of which sheet then

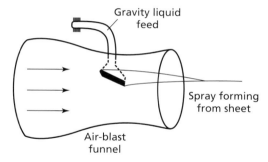

Gravity liquid feed

Spray forming from sheet

Air-blast funnel

Fig. 4.8. Venturi nozzle (on the principle of a scent spray).

fragments into fine drops. Vmd values are therefore small, but so are nmds and the fine mist quickly loses its momentum and drifts round rather than onto targets.

Foggers

Oil solutions or emulsions of insecticides can be introduced into a cold high-velocity air stream generated by spraying machines to produce insecticidal fogs or aerosols. Alternatively the insecticide can be introduced into specially heated chambers or into hot exhaust fumes of spraying machines to produce thermal fogs or aerosols. Care may be needed with thermal foggers to ensure they do not become flame-throwers! Usually the term fog is applied to formulations composed of droplets with a vmd of < 50 μm while mists comprise droplets of about 50–80 μm. Fogging machines can be portable, mounted on vehicles or on aeroplanes and helicopters, or as permanent installations in glasshouses.

Ultra-low-volume (ULV)

The main bulk of any liquid insecticidal formulation consists of the diluent or solvent; the actual amount of insecticidal active ingredient is small. Greater efficiency can be achieved if concentrated insecticidal solutions are sprayed sparsely over an area, than if much larger applications containing the same quantity of insecticide but dispersed in a large volume of solvent are used. With ULV spraying techniques application rates may be 75 ml to 1 litre/ha instead of 5–100 litres or more.

Fig. 4.9. The 'Micronair' spinning cage set up for hand-held application.

Spinning cage

This development, which has led to improved droplet spectra, arose from the need for spray equipment for fixed-wing aircraft to have a high output per minute if an adequate dose per hectare is to be achieved at ground speeds of about 100–150 km/h. A typical 100 litres/min output requires about 24 traditional nozzles on the aircraft's boom, but four spinning cages can achieve the same output. The spinning cage (Fig. 4.9) involves the spraying of pesticide through a diffuser tube onto a metal gauze cage spun rapidly by small propellers rotated by the airflow across the wing. This use of centrifugal rather than hydraulic forces to produce droplets results in a still fairly wide droplet size spectrum (vmd/nmd ratio is about 3.5), but without the production of the myriads of tiny drops or the few really large drops so wasteful of insecticide that is characteristic of hydraulic nozzles. In fact, the restricted droplet spectrum of the spinning cage means that flow rates as high as 100 litres/min are not necessary in practice.

Spinning cup

This second development towards narrower droplet spectra stemmed from attempts to provide the peasant farmer in developing countries with cheap handheld spraying equipment using very little water. Water, often a scarce commodity during the crop season in the tropics, may also need to be carried a large distance. Contemplating reducing applied volumes to perhaps only 1 litre/ha forced a new approach to the production of droplets, since, clearly, large drops were wasteful and had to be eliminated from the spectrum. To break up 1 litre to cover 1 hectare effectively leaves little margin for any wastage; it is equivalent to distributing the contents of a wine bottle evenly over the area of two football pitches! The machine eventually developed, the spinning cup (Fig. 4.10), involved the break-up of rods of liquid rather than sheets. This is achieved by allowing liquid to dribble under gravity on to the inside of a grooved plastic cup spun rapidly by a battery-powered electric motor. The liquid is spun outwards by centrifugal force to form rods along the fine grooves machined in the face of the cup. When these rods disintegrate into droplets as they leave the cup, a spray with an extremely narrow droplet spectrum can be produced, and the desired droplet size can be varied over a considerable range by changing the speed of rotation of the cup. Output is, however, limited by the flow rate the cup can take without flooding and forming a sheet of liquid around the edge. However, it has been possible to increase the output at each 'nozzle' position by using much larger cups, stacking several cups on one spindle and by an ingenious use of a smaller cup inverted inside a larger one. With such solutions, spinning cups have been used on tractor-mounted equipment to spray field crops and orchards.

The vmd/nmd ratio of the spray is around 1.3. The production of such a narrow spectrum is known as 'controlled droplet application' (CDA). Under the right environmental conditions and in skilled hands, the 10 g of active ingredient on plants which seems to be needed with many compounds to give effective pest control can be achieved with as little as 30 g/ha issuing from the spinning cup. With quantities as low as 0.5 litres/ha now practical, oil rather than water can be used as the carrier to increase the weight and decrease the evaporation of small drops. In theory, the spinning cup is a great stride forward; if drops, because of their more uniform size, can be

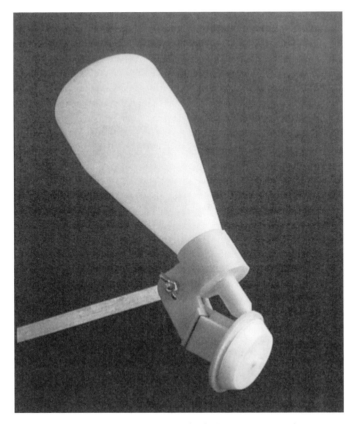

Fig. 4.10. A spinning cup nozzle (Micron sprayers).

deposited on plants in large proportion to the total volume applied, it must be possible to use less actual pesticide per hectare. Unfortunately, in practice, the reduction in spray volume obtainable with these devices has not been accompanied by recommendations to reduce the rate of insecticide; indeed, attempts to reduce the rate have frequently led to failure of pest control. The reason is that with such a narrow droplet spectrum, if the droplet size is not perfectly matched to the target and the environmental conditions, droplet capture will be very poor. Getting it right requires an often impracticable level of sophisticated measurement of environmental variables and subsequent interpretation. Perhaps the spinning cup has perfected a technology of droplet production ahead of our ability to exploit it. The older hydraulic sprayers, with their wide droplet spectrum, at least have a built-in reliability factor. Under very different crop and environmental conditions, the 750 g of insecticide sprayed will always have about 10 g distributed at the right droplet size, whatever the conditions that day. The spinning cage is perhaps as far as we can go with confidence at present in reducing the spread of the droplet spectrum.

Electrostatic sprayers

This third development, another form of CDA with a narrow droplet spectrum (vmd/nmd ratio as low as 1.1), originated with the same motive as that which inspired the development of the spinning cup, i.e. to reduce dependence on water as a carrier needed in large volume. The most successful such sprayer was probably the 'Electrodyne', into which liquid was fed by gravity between two parts of a nozzle charged at 25 kV. The voltage between the nozzle and an earthed electrode formed an electric field which both atomized the ligaments and charged the droplets. The charged spray cloud induced the opposite charge on the target, and became attracted to that target giving good 'wrap-around' of both upper and lower leaf surfaces of isolated plants. This enabled very small drops of 40 μm, which would otherwise drift, to be produced, but only certain oil formulations could be used, though the low volatility of the oil contributed to the minimized drift. Volumes as little as 0.33 litres/ha could be applied. Penetration of crop canopies was rather poor, however, since the spray was attracted to the part of the plant nearest the spray. However, this characteristic could have had a particular advantage in pest management (see Chapter 13). Also, any crosswind deflected the droplets elsewhere.

The company that developed the electrostatic sprayer has now decided that the problems of the sprayer make its uptake unlikely, and the project has been abandoned. It does seem a pity to lose this superb way of making uniform small drops without the need for moving parts in the machine.

Twin-orifice sprayers

While the spinning disc and the electrostatic sprayer were being developed in Europe, a research scientist in a remote part of China, without recourse to the European literature, identified drop size variation as a problem crying out for a CDA solution. His solution was quite different from either of the European approaches. He modified an ordinary knapsack sprayer so that pumping the handle pumped both liquid and compressed air in separate tubes, the liquid tube running inside the air tube. The low pressure produced by air over the liquid orifice draws off droplets of good uniformity. The same principle is used with the simple Canadian Ogee nozzle – the liquid at the orifice of this spade-shaped nozzle appears to boil with the reduced pressure of a powerful air-blast passing across it.

Air assistance

The concept that there is a single correct drop size on any one occasion is over-simplistic. Out-of-doors the drop will encounter different speeds and directions of air movement between leaving the nozzle and impacting on a leaf in the crop canopy. Each stage of the journey would require a different 'ideal' size of drop. This of course lies at the heart of the difficulties people have experienced trying to bring dose rates down with the spinning cup and its very uniform-sized drops. The movement of a drop will be the balance between three vectors – upward thermal currents, downward gravity and lateral wind. Whatever the balance when the drop leaves the nozzle, not only will the drop immediately begin to reduce its size by evaporation, but the vectors will change dramatically as the crop surface is reached. Here lateral movement will be reduced and upward movement increased by air turbulence caused by the irregular height of different plants. If successfully penetrating this barrier to downward movement, gravity will suddenly become more important and lateral air movement greatly reduced within the crop.

So what can be done? The answer is to use a fan to provide sufficient air movement to override the ambient vectors. Increasingly this approach is being taken. Orchard sprayers with hollow-cone nozzles have for many years used air from a large fan behind the nozzles to push the droplets into the canopies of orchards, and air-assisted spinning cup sprayers are now available. The spinning cage always relies on air movement to push droplets into the crop, either the downwash from the aeroplane or, in hand-held or ground machine-mounted versions, from the propellers (which rotate the cage with the airflow over an aircraft's wing) being rotated by the motor which spins the cage or by an air-blast from a motor in a mistblower version. Recently the first air-assisted tractor-mounted boom sprayers have appeared with a huge central fan distributing air over each nozzle along trunking over the boom. To see such machines running with and without the fan is a convincing demonstration of the value of air assistance. Also, the air 'bouncing' back off the ground improves the deposition of drops on the underside of leaves.

The use of air to deposit the drops produced by the electrostatic sprayer could solve the problems that this machine has shown. Paradoxically, the very feature the machine was designed for, to produce charged drops, really needs overriding to get penetration into a crop. Yet electrostatics is a wonderful way of producing very even-sized small drops with a very cheap machine, and it seems a great pity not to exploit this principle and develop the electrostatic sprayer further.

The only development which is perhaps still generally needed is to achieve some spatial separation between droplet production and the air propulsion. This would be akin to a tennis serve and would prevent the air-blast forces contributing to droplet production.

Pressurized aerosol dispensers

Aerosols are produced from pressurized cans filled with a gas that acts as a propellant and which on release vaporizes leaving the insecticide, often a synthetic pyrethroid, dispersed as fine droplets. Such aerosol dispensers are used on aircraft to kill exotic insects in the cabins and often by individuals to kill a range of domestic insect pests. They are often used by home gardeners but, in outside conditions, the small drops produced are very subject to drift. Because of increasing concern about the deleterious effects on the ozone layer of the propellant gas, efforts are being made to find a harmless propellant or to replace these aerosols by simple pump-action sprayers.

Film-coating

This is a relatively new but potentially important technique, whereby seeds (such as the relatively large seeds of brassicas) are passed through a fine spray of recirculating insecticide in special chambers. The spray leaves a fine film of insecticide on the seeds, which protects them and the seedling roots against soil pests. Since this enables very low doses (e.g. only 2 g/ha for organophosphate insecticides) to be effective, the technique has enormous potential for safety for natural enemies (see Chapter 13) and for reducing environmental contamination.

Application of solids

In developing countries dusts are unfortunately often applied manually and very simply. They may simply be broadcast by hand from a container, or distributed by tapping a mesh dust-filled bag, often a nylon stocking. Both methods are likely to result in considerable contamination of the person applying the material. They are best applied on crops early in the morning,

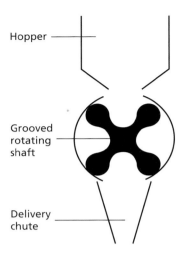

Hopper

Grooved
rotating
shaft

Delivery
chute

Fig. 4.11. Principle of granule applicator.

when the dew on the leaves enables the dust to 'stick'. Dusts formulated with stickers for seed dressing are stuck to seeds in a rotating tumbler.

Dusts are rarely applied by machine, though there are applicators suitable for distributing both dusts and granules. For example, a mistblower can be adapted to produce a stream of dry material. Alternatively, simple bellows-type dusters can be used, e.g. to dust peoples' clothing with insecticides to control body lice (*Pediculus humanus*).

Wheeled applicators for dry formulations may be pushed by hand or mounted on a tractor; they are more usually used for applying granules than dusts. The basic feature is a grooved shaft (Fig. 4.11) rotated by a wheel moving over the ground. The grooves are fed by gravity from a hopper holding the granules/dust, and emptied as the groove rotates 180° to point downwards. In tractor-mounted machines, an air-blast may assist even distribution of the granules. Another variation is that the material is distributed by pipes to a small box mounted above the tines of a seed-drill; this enables granules to be deposited in the row at the time the seed is drilled.

Special forms of application of insecticide

Smokes

When lit, small cardboard smoke generators (containing insecticide mixed with a pyrotechnic powder as used in fireworks) will burn for a few minutes

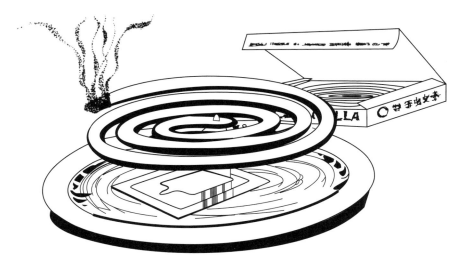

Fig. 4.12. Mosquito coil containing pyrethrum or a pyrethroid which on burning provides a mosquito repellent smoke. (From Rozendaal, J. A., 1997).

to produce an insecticidal smoke. This fumigates enclosed spaces, such as bedrooms to kill bedbugs (*Cimex* species), or glasshouse pests hiding in inaccessible places or wood-eating beetle infestations in attics. Similarly, mosquito coils (which will also kill other biting insects) when lit burn to release an insecticide (usually a pyrethroid), but the idea here is for them to burn slowly, such as throughout the night (Fig. 4.12). Simple coils consist of combustible material, an insecticide, usually pyrethrum or a pyrethroid such as allethrin or esobiothrin, and sometimes a scent is added to make them more acceptable when burnt. When lit one coil produces sufficient insecticidal smoke to repel mosquitoes and other indoor biting insects in a small room (about 35 m^3) for most of the night. Locally made coils are often of poor quality and ineffective because they contain too little insecticide. Where electricity is available more sophisticated vaporizing mats can be used which have the advantage of not producing smoke. The insecticidal mat (about 35 × 22 × 2 mm) is usually made of compressed porous paper impregnated with allethrin, bioallethrin or esobiothrin and is placed on an electrical heater to vaporize the insecticide. The blue-coloured mat changes to white as the insecticide evaporates and is used up. One mat effectively repels, and sometimes kills, mosquitoes for up to 8–10 hours; in larger rooms two or more mats are needed. An even more sophisticated, but more costly, method heats a wick placed in a reservoir bottle containing a liquid insecticide. One bottle can provide 8–10-hour protection for about 45 nights.

Fumigation

Highly volatile and usually also highly toxic compounds such as methyl bromide (now banned in the UK) have been used to sterilize bulk materials such as soil and piles of stored products such as grain. Compounds used for fumigation are usually very broad-spectrum and kill nematodes and fungal pathogens as well as insects. Thus soil fumigation tends to give higher yield increases in the next crop to be grown than can be accounted for by insect control. Because of the dangers the materials pose to operators, fumigation is very often an expensive exercise carried out by specialist contractors basing their dosing on careful calculations of volume to be treated, gaseous diffusion and residual life of the toxin. The fumigant (normally a liquid) is pumped in under plastic sheeting to contain the liberated gas, and in piles of stored products, fans are run under the sheet to ensure dispersion of the gas through the stack.

Phosphine (aluminium phosphide) tablets have been added to bulk containers of food that are transported by sea to release a poison gas and so protect the cargo during a long journey. The dose has to be carefully calculated so that the gas has broken down by the time the container is opened at its destination, and there have been human fatalities at ports when the wrong dose of fumigant has been added to the container.

Spot-on and pour-on applications

It has frequently been recommended that sprays should be applied only where infestation is seen – so-called spot applications, which are possible on a small scale. Some years ago, in order to protect ground beetles predating the eggs of cabbage root fly (*Delia radicum*), special spraying machines were developed which only produced a burst of liquid when the stem of a plant interrupted a light beam.

Occasionally focal topical applications of insecticide are used on animals. For example, solutions of permethrin or fenthion can be applied to the skin of cats and dogs in just one small area. The insecticide is absorbed through the skin and passes to the animal's blood, so that blood-feeding arthropods ingest the insecticide and are killed. A single treatment lasts for about 3–4 weeks. Insect growth regulators such as methoprene or pyriproxyfen can also be applied topically to give prolonged ovicidal activity. Also when IGRs such as cyromazine and lufenuron are administered orally once a month, arthropods feeding on cats or dogs produce non-viable eggs for several weeks. On a larger scale, specially formulated solutions of insecticides are poured along the backs of livestock, and aided by surfactants in the solution the toxicant disperses over hair and skin of an animal spreading rapidly to most parts of the body.

Self applications

An example is the treatment of posts and other structures of self feed-lots with insecticides, such as amitraz, so that when livestock feed they automatically rub their heads, necks and other body parts against these treated surfaces. Clearly this does not give as much control as spraying cattle or using pour-on insecticides. A better system is the so-called '4-poster' topical treatment device which has four upright vertically mounted rollers into which amitraz is automatically fed from a reservoir which results in livestock treating themselves with a pour-on insecticide when they feed.

Impregnated protective materials

Ronald Ross, proponent of mosquito nets in the early 1900s, would have been intrigued to learn that bed nets, albeit insecticide-impregnated ones (Fig. 12.1), came back into favour in the 1980s.

Many bed nets used by poor communities are cheap, badly made, easily tear and so become useless in affording protection from bites by mosquitoes and phlebotomine sand flies (vectors of leishmaniasis), triatomine bugs (vectors of Chagas disease) and bedbugs. Also, vectors will readily bite through a net if any part of the body is pressed against it (see Chapter 12). Nets impregnated with pyrethroid insecticides such as permethrin or deltamethrin, which repel and kill mosquitoes, will still protect people against bites even if nets are torn or people sleep up against them. Nets can be impregnated by simply dipping them in insecticide contained in a plastic dustbin, a task that can be carried out by community health workers or by net owners. Such treatment can remain effective for six months, or for a year or more if lambdacyhalothrin or alphacypermethrin are used.

In Vietnam and China, where more than 2.5 million households are protected by impregnated nets, there have been significant reductions in malaria transmission. It seems that, at least in some circumstances, use of impregnated nets can have a mass killing effect, and thus protect those not sleeping under nets. Clearly bed nets will not give protection against vectors that bite people during the day or early evening before people have gone to bed.

Baits

These incorporate an insecticide (solid or liquid) into food that is highly attractive to various arthropods. Such toxic baits can remain effective in attracting and killing cockroaches and house flies and related flies for many weeks. Baits incorporating molluscicides such as metaldehyde are also available to kill slugs and snails, and insecticide sprays may incorporate attractants to improve their effectiveness (see Chapter 13).

Targets

Targets, or screens, can (especially with baits or utilizing insect species-specific behaviours) be selective and, if treated with insecticide, restrict how far the environment is contaminated. Insecticide-treated targets, screens and traps are increasingly being used to kill tsetse flies (vectors of sleeping sickness and animal trypanosomiasis). Tsetse flies are attracted to dark colours, especially to black or dark blue, so targets made of dark-coloured cloth and often incorporating an attractant odour such as 1-octenol-3-ol, acetone or butanone can be mounted between two upright stakes and sprayed with pyrethroids. This provides an environmentally acceptable way of attracting and killing the flies. Some 50 000 odour-baited targets are providing good control of tsetse flies in Zimbabwe.

Somewhat similarly, cards or cords soaked in insecticides, such as diazinon or permethrin can give effective and environmentally safe control of house flies and related Diptera in cow sheds. Such cords, however, should be dyed red to alert people that they are insecticidal.

House spraying

Aimed at killing resting vectors, especially mosquitoes, the practice involves using a pressurized sprayer to spray water-dispersible (wettable) powders of residual insecticides on the interior walls and ceilings or roofs of houses (Fig. 4.13). The effectiveness of indoor spraying clearly depends on the arthropods resting in houses, and many important malaria vectors (*Anopheles* species), especially those in South-East Asia and Latin America, feed and/or rest out of doors, and are thus not killed by indoor spraying. There is also evidence that house spraying may either promote the selection of exophilic (outdoor resting) populations of a particular vector species, or favour the increase in numbers of a sibling species that rests out of doors while reducing

Fig. 4.13. Spraying inside a house for malaria control (Courtesy of the Wellcome Trust and the late L. J. Bruce-Chwatt).

populations of a species of the complex that is primarily indoor resting. Such population changes have been recorded in the *Anopheles gambiae* complex in Zimbabwe, and in other vectors such as in Venezuela and Thailand. The end result can be that malaria originally transmitted by indoor-resting mosquitoes is now maintained by increased populations of an outdoor-resting vector.

DDT has been the main chemical used for house-spraying until recently (see p. 109). A well known disadvantage of DDT is that it can be an irritant and cause insects to leave sprayed surfaces before they have picked up a lethal dose of insecticide. To a lesser extent this also applies to the synthetic pyrethroids, which have often replaced DDT where its use is no longer permitted.

Insecticide application to aquatic habitats

This may take the form of spraying such habitats (Fig. 4.14) to kill mosquito larvae, or of adding insecticide directly to rivers to kill black fly larvae. Probably more effort and money is expended in the USA on mosquito control, mainly larviciding, than anywhere else. Widespread spraying can lead to ecological problems (see Chapter 5), especially as there is no residual effect and spraying in tropical areas may need repeating as frequently as every 10–14 days. In the past

Fig. 4.14. Spraying a mosquito larval habitat with an organophosphate insecticide (M. W. Service).

aquatic waters were dosed with organochlorines, but today the less damaging organophosphates, carbamates and pyrethroids are used. Nevertheless, none of these insecticides is specific and varying numbers of non-target organisms will be killed, including in some instances fish. In Kenya, when rice fields were sprayed against rice borers, this resulted in an increase in mosquito populations (see p. 113). However, many mosquito larval habitats are very small and often transient (e.g. puddles, small pools, ditches) and are not colonized by fish or even by many invertebrates, so ecological damage is minimal.

Direct dosing into rivers has been especially targeted at the immature stages of simuliid black flies, which are vectors of river blindness (onchocerciasis), and are restricted to flowing waters, ranging from small streams to the largest rivers.

Because of the severity of river blindness in the Volta River Basin area of West Africa and its devastating effect on rural life, the world's largest and most ambitious vector control programme was initiated by the World Health Organization in 1974. This was such a landmark in medical entomology that it merits the following detailed account.

The programme was called the Onchocerciasis Control Programme, universally known as OCP. Originally the scheme covered seven countries, but

Fig. 4.15. Helicopter dropping temephos insecticide into a stream to kill larvae of simuliid black flies (Courtesy of J. B. Davies).

it now covers Benin, Burkina Faso, Côte d'Ivoire, Ghana, Mali, Niger, Togo, parts of Guinea, Guinea Bissau, Senegal and Sierra Leone. Some 50 000 km of rivers over an area of 1.3 million km² that were breeding sites of the simuliid vector (*Simulium damnosum*) were dosed at weekly intervals with temephos. The insecticide was dropped from helicopters or fixed wing aircraft (Fig. 4.15) at pre-selected sites and the water carried the insecticide downstream killing the larvae as it passed. Temephos resistance appeared in some areas in 1980, so in these sites the microbial insecticide *Bacillus thuringiensis* var. *israelensis* (or the preferred WHO name of *B. thuringiensis* serotype H14 because it lacks political overtones!) was used. Larviciding needed to continue until the year 2002 in countries that were added later, because the parasite reservoir in humans can live for 20 years, and if flies were present they would become infected and transmit onchocerciasis. Since 1988 the drug ivermectin (see p. 68)

Fig. 4.16. Cows being dipped in an insecticidal bath (Courtesy of the Food and Agriculture Organization of the United Nations).

has been given orally once or twice a year to kill the microfilarial parasites in the human population.

The programme has been spectacularly successful, with transmission ceasing over much of the OCP area.

The chief insecticide used, temephos, has undoubtedly killed non-target aquatic invertebrates as well as some fish, but there has apparently been no serious ecological damage to river fauna.

Treatment of livestock

Application of insecticides by dipping or spraying remains an important method of controlling ticks on livestock. Originally organochlorines were used, but absorption into lipids created residue problems in meat.

Fig. 4.17.   A cow going through an insecticidal spray-race (Courtesy of the Food and Agriculture Organization of the United Nations).

Organophosphates, which are rapidly detoxified and eliminated from live-stock, replaced the organochlorines, but now there is mounting evidence that exposure of farmers (who can easily be heavily contaminated when controlling the animals through the dip) carries considerable risk. Carbamates and pyrethroids have therefore been increasingly used.

Dipping cattle in an insecticidal bath (Fig. 4.16) is more reliable than spraying; it is simple, ensures good coverage and not much can go wrong, whereas mechanized spraying (Fig. 4.17) or even hand spraying can be unreliable, prone to failure and give poor coverage of the animal. On the other hand spraying equipment is often cheaper than building a dip and can be portable. If there are tick-resistant cattle, dipping often needs to be repeated only 2–3 times a year, but in many developing countries weekly dipping is routine.

Restricting insecticidal spraying to just a few selected sites on cattle, e.g. the ano-genital area, udder regions and the axillae reduces costs but tick control is not so effective.

Deposits and residues

The deposit is formed by that fraction of the applied insecticide which both contacts and then adheres to the treated surface. Additives such as 'wetters' and 'spreaders' are added to the pesticide to improve the 'flattening' and retention of the drop on impact with the leaf or the wall of a house. Too much wetter will result in the drops coalescing and liquid running off the leaf. The correct amount of additive will vary with individual crops and their leaf surface properties; the lowest amount of additive, which gives adequate wetting, will also give the maximum deposit of pesticide. When house walls are sprayed, spray men must be trained to stand the correct distance from walls and move the spraying lance at the correct speed to avoid 'run-off' and to ensure the correct dosage over large uniform areas.

The point was made earlier that there is a huge loss of insecticide between release at the nozzle and deposition on the plant. This loss continues, for the deposit is almost instantaneously changed to a residue as loss of pesticide from a variety of processes begins.

This degradation of the pesticide follows a hollow logarithmic curve, not a straight line. The persistence of residues is measured in terms of 'half-life', and it is important to understand what this means. Half-life is the time taken for residue levels to fall by half. There are thus more than two half-lives to the life of a residue. When we say the half-life of dieldrin is 35 years, this means that every 35 years what is there will be halved. So after 35 years half of what was applied will remain, and after the next 35 years half of that half (i.e. a quarter of the original amount) will still be there. So 105 years after application, a sixteenth of the amount applied will still remain.

Water evaporating from the leaf surface will carry dissolved volatiles away, and falling rain will wash the residue off surfaces to a greater extent the more a 'wetter' has been used, whereas 'stickers' will increase retention. Wind-borne dust abrades the layer of insecticide; leaf growth and daily expansion and contraction will cause brittle dry deposits to flake off. When people sit up against sprayed house walls, much of the insecticide deposit is rubbed off, and because some disease vectors rest mainly on the lower parts of walls this can present a problem. The deposit is also very vulnerable to chemical degradation; some of this is enzymatic within the plant or by the leaf surface microflora, but especially important is oxidation, usually catalysed by the ultraviolet radiation in sunshine ('photochemical oxidation').

Much of what remains of the 'residue' will never be contacted by an insect. Even the small proportion of the total residue on the crop picked up by insects may partly be knocked off again as the insects clean themselves or move it

Table 4.2. *Toxicity to* Anopheles *of DDT applied at various doses to mud blocks (in each column, the highest value is picked out in bold type). From Hadaway and Barlow (1951).*

Dose (mg/ft^2)	Crystal size (μ)	DDT picked up per fly (μg)	% kill (in 24 h)	Kill per μg picked up
0.75	0–10	0.13	55	**423**
6.00	10–20	1.00	**100**	100
50.00	20–40	**1.70**	23	13
400.00	**40–60**	0.25	0	0

to parts of the body (e.g. the wings for many insects) where it will have little effect. The importance of where the insecticide finishes up is clearly made by some data on residues accumulated by flying as opposed to resting insects.

Flying locusts in a space spray accumulated twice as much insecticide as resting ones, because the wing beat pulls in the small drops. However, since insecticide accumulating on the wings has no toxic effect, over four times the amount of insecticide needed to kill the resting locust was needed to kill the flying ones.

Most people assume that the greater the dose, the greater the residue and therefore greater the kill of the insect. However, the interaction of the living insect with the non-living residue can cause such correlations to fail. This is beautifully illustrated by some early work (Table 4.2) by Hadaway and Barlow (1951), using DDT against an *Anopheles* mosquito on surfaces mimicking mud walls of houses. Hadaway and Barlow applied an enormous range of doses, the highest dose being over 400 times the lowest. The size of the particles of the deposit was of course correlated with dose. However, large crystals picked up by the insect get brushed off again by its cleaning movements, so that the amount of insecticide remaining on the insect is not positively correlated with dose rate. Furthermore, small particles are more efficient than larger ones at killing the insect. This is because the area of contact of a given amount of insecticide with the cuticle, which largely determines the rate of diffusion into the insect, is greater if the insecticide is in many small particles and not a few large ones. They and other co-workers also found the persistence of insecticide wettable powders varied according to the different types of mud used in house construction. In fact they had different types of mud flown from Africa to the UK for experiments on pesticide persistence.

All this leaves an infinitesimal fraction of the pesticide that was bought on a useful part of the surface of the cuticle of insects. This minute fraction still has to be adequate to sustain the further losses that yet have to occur, and to accumulate in sufficient quantity at the site of action within the insect. The first barrier is the cuticle. Not surprisingly, this external skeleton of the insect is adapted to insulate the insect from the external environment. Insects, as terrestrial creatures, have to protect themselves against water loss; the outer layers of the cuticle, particularly the thin surface wax layer, form an effective barrier to the movement of water. The tissues and body fluids of insects contain many non-aqueous compounds, and the inner layers of the cuticle resist the penetration of such compounds, including oils and waxes. Thus the cuticle can resist penetration by insecticides which are solely soluble in either water or oil. An effective insecticide therefore should have solubility in both oil and water (a good oil : water partition coefficient) and thus be able to move in either phase into different parts of the cuticle. This can be achieved by choosing a suitable solvent.

The site of action of an insecticide is in a specific tissue (usually an enzyme in that tissue) representing a small fraction of the total insect and surrounded by a mass of physiological and biochemical barriers or alternative sinks for the toxic molecule. Many of the non-target tissues (e.g. fat) will store insecticide, and many tissues, including body fluids, carry enzymes capable of oxidizing, hydrolysing or otherwise splitting the toxic molecule. Insects, particularly plant-feeders, need such a powerful armoury of enzymes to deal with the many natural toxins they encounter.

Killing an insect is like trying to fill a bucket full of holes. It is not the amount of insecticide which reaches the target site that is critical; it is its accumulation, i.e. the rate of arrival minus the rate of removal. Thus many times the actual lethal dose must be picked up on the cuticle of the insect for death to follow. The main problems are passage through the cuticle at a sufficient rate, storage in the cuticle and fat, slow diffusion through the skin underlying the cuticle and through the other tissues, excretion, lack of penetration of the sheathing of the nerves (often the site of action), which is a barrier to ionized materials, and the breakdown of the toxin by insect enzymes. The last-named is particularly relevant to the insect's resistance to insecticides (see Chapter 5).

5

Problems with insecticides

Introduction

The first synthetic broad-spectrum insecticides, the organochlorines, offered a revolution in the efficacy of pest control, especially of the malaria mosquitoes and other disease vectors of mankind. Additionally they offered cheap, sure and long-lasting control of crop pests. With hindsight, we can be critical of the profligate early use of organochlorine insecticides: however, we need to remember that it is hard to predict the unforeseen. Many of man's technical advances have in the past needed, and will continue to need, modification in the light of experience. Progress will always involve risk. The wide-scale use of the early synthetic broad-spectrum insecticides, particularly certain organochlorines with the then prized characteristic of long persistence, produced certain obviously damaging side-effects. The scientific community, operating via committee structures and accurate but rather conservative memoranda through official channels, was drawing the attention of governments to these problems as early as the mid 1940s. However, the real landmark in change of attitudes and government legislation was the publication of Rachel Carson's *Silent Spring* in America in 1962 (Fig. 5.1). This castigation of insecticides was considerably overstated in the opinion of many scientists, but it created a public awareness and outcry of which politicians could not help but take notice. Her main criticisms, in order of the importance she gave them in the book, were the toxicity of pesticides to humans, their damaging effects on wildlife and lastly, the appearance of pest strains with tolerance to the toxins. Today, we would probably completely reverse Miss Carson's order of importance.

Fig. 5.1. The cover of the paperback edition of *Silent Spring*, with the emotive image of a dead bird. Note in the top right corner that the price of the book in 1962 was five shillings (25p or 35 US cents).

Before looking at the problems identified by Rachel Carson, it is important to point out that they may not occur in isolation, and that politics and human behaviour can cause problems with the effectiveness of insecticides just as much as biological phenomena. The classic case is the failure of the eradication of malaria.

Case history for lessons in failure: malaria eradication

Prior to the DDT-era malaria control was based on mosquito larval control and sometimes also the administration of the antimalarial drug quinine. In addition to repeatedly spraying larval habitats with oils and Paris Green (about every 10–14 days) habitats were also frequently filled in, drained or otherwise eradicated. When DDT became available the control strategy switched to killing adult mosquitoes that rested prior to, and after, blood-feeding in houses. DDT wettable powder was sprayed on the interior walls, ceilings or roofs of houses. A deposit of 2 g/m^2 remained effective for about six months. For the first time this meant that it was logistically possible to spray houses over very large areas, even countrywide. Optimism was so high that the WHO declared in 1955 that global malaria eradication was possible, albeit it would be difficult in sub-Saharan Africa because transmission was so intense, and there was also a sense of urgency because problems of insecticide resistance were recognized.

What were the results? Well, malaria was eradicated from Europe, the USA, northern Australia, Taiwan, the Seychelles, Mauritius and some Caribbean islands. But this was relatively easy because either small islands were involved or malaria was at the edge of its distribution where transmission was not intense. So what about malaria in the tropics? In India there were an estimated 47 million cases in 1947. In 1958 India embarked on a programme of malaria eradication based on spraying houses with residual insecticides, such as DDT. In 1965 malaria cases were just 0.1 million; clearly malaria eradication was within reach. Then in 1977 there were an estimated 7–10 million cases. No one knows the true figures. What went wrong? It is true that malaria vectors became resistant to DDT and other more expensive chemicals had to be used, but it appears that socio-economic reasons were to blame. International funding was reduced at a time when more expensive insecticides were being used. Poor parasitological and entomological surveillance resulted in local malaria outbreaks not being identified and dealt with rapidly. Also, the dull repetitive routine of spraying made it increasingly inefficient; householders began to refuse spraying because it made their walls messy and it was inconvenient. Interestingly people initially welcomed the spray men because, not only were mosquitoes killed but also other pests such as bedbugs, but bedbugs became resistant and their populations increased, so the people then blamed spraying for an increase in bedbugs. The problem is that spraying has to continue, even though transmission may be greatly reduced, until eradication has been achieved, and this has proved unsustainable.

Other countries faced problems in trying to eradicate malaria, including the recognition that not all malaria vectors rest inside houses, and consequently they escape the effects of residual house-spraying.

In 1969 the WHO was forced to abandon its dream of eradication, and accept that at best only malaria control could be achieved in most countries. Since then the emphasis has been on an integrated approach involving use of insecticide-impregnated mosquito nets, drugs and vaccines. Nevertheless, whereas malaria was endemic in some 143 countries or areas, 37 have been freed of malaria while in another 16 countries transmission has been substantially reduced. Nevertheless that leaves 90 countries still having moderate to high levels of transmission. There has been virtually no impact in malaria transmission in tropical Africa, where about 90% of all malaria, and 80% of all deaths, occur, including an estimated one million infant deaths a year.

Toxicity to humans

Because of the very rapid kill that nerve poisons can achieve, insecticide development has largely focused on molecules with nerve-poison properties. Unfortunately, there is considerable similarity between the biochemistry of the nerve system between insects and all other animals, including man. So the dilution of the concentrate and the spraying of poisonous chemicals present an undoubted hazard to the person applying the chemical. Rough estimates based on WHO figures suggest that in 1973 there were perhaps 500 000 cases of such occupational accidental poisoning by pesticides, though only an estimated 1% of the cases resulted in death in countries where suitable medical treatment and antidotes were readily available. In 1985 there were an estimated 3 million hospitalized poisonings and about 220 000 deaths, and in 1999 WHO estimated there were some 257 000 poisonings.

Insecticides are likely to have been responsible for the largest proportion of these deaths, though many fatalities have also involved fungicides and especially herbicides. Such fatality figures compare well with the industrial risks of many other human occupations. Accidents are often the result of ignoring safety procedures, which have been established for handling pesticides just as for other industrial operations. It is, of course, sometimes hard to reconcile safety procedures with practice; for example 'fully protective clothing' may be both too expensive and uncomfortable for a peasant farmer or a public health spray man in a tropical climate. Manufacturers do take great care to get the safety message across. Where language may be a problem, safety instructions are translated into cartoons.

When people sprayed houses with DDT in anti-malaria campaigns, they commonly ignored safety regulations, but because of the very low toxicity of DDT to humans, they got away with it and suffered no adverse effects. However, when DDT was banned for political reasons, or DDT resistance appeared in the malaria vectors, they began spraying the ecologically more acceptable organophosphates, and their disregard for safety regulations made many succumb to acute insecticidal poisoning, and sometimes death. Similarly, numbers of farmers dying from acute insecticide poisoning have risen sharply in countries like Sri Lanka as soon as DDT was banned and replaced with the organophosphates with their far greater toxicity to humans, in part stemming from their great volatility in hot climates.

Other fatalities arise from ignorance or negligence, such as storing surplus insecticide in used soft drink bottles or using insecticide containers for storing water. Other fatalities represent murder and suicide. In these respects insecticides are just as prone to misuse as other toxic chemicals (aspirin, bleach, disinfectant etc.) regularly stored in the home.

A concern more specific to pesticides is any danger attached to the chronic regular intake of small quantities as residues in our food. As mentioned earlier, like radioactivity, pesticide levels decrease with time along a hollow curve to approach zero asymptotically. There is therefore theoretically no time at which the residue reaches zero. In the 1950s, 'zero' was when the analytical equipment then available could no longer detect any traces of the poison. The equipment of today can detect traces below one part per 1000 million (a thimble-full in an Olympic size swimming pool), and to aim for 'zero' residues is no longer an option. We have to live with the fact that we are exposed to insecticide residues, however small, in our food.

Great alarm followed the discovery in the 1940s that post-mortems of, for example, accident victims revealed measurable quantities of DDT in the body fat of nearly all inhabitants of developed countries and that no restaurant meal could be found free of DDT residues. The alarm was heightened by the linear increase of these DDT residues in body fat year after year. However, it could later be shown that man eventually attained a plateau level of about 10 parts per million (ppm) DDT in body fat, and that thereafter further intake was balanced by elimination from the body. Modern pesticides are much more quickly broken down and lost from our bodies. Nevertheless, it is still reasonable to ask 'Do such residues do us any harm?' It is obviously impossible to guarantee that they do not, and there has been no shortage of effort to try and demonstrate that they do. The failure of such efforts in spite of the fact that man has been imbibing insecticide residues for 60 years does suggest the risk is very small when compared with the improvements in health and in food supplies that insecticides have achieved for us. However, alternative ways of protecting our crops or at least reducing the amount of pesticide used are worth serious consideration if they have the potential to reduce the risk still further.

Effects on wildlife

One of the most publicized side-effects of organochlorine pesticides was the death of many birds. The Penguin edition of *Silent Spring* makes this point with its colour picture of a dead bird on the cover (Fig. 5.1). This death of birds had two main causes. First, the compounds were being used widely as seed dressings for grain which was then eaten by birds; second, pesticide applied to crops by drenching the soil leached into waterways, where it was concentrated in particular water layers, and through food chains until the higher predators became affected. Predatory land birds such as eagles, kestrels and hawks as well as water fowl, particularly grebes, all suffered from organochlorine poisoning. Some of this poisoning was acute, but in the case of some of the predatory birds, post-mortem analysis revealed organochlorine residues well below those which would have killed them. It seems likely that sublethal effects of residues can cause death by affecting the behaviour (in the case of the raptors, their hunting skills) of the victims.

There has in the past sometimes been uproar among villagers over death of their chickens which had eaten dead and dying flies, ticks and cockroaches contaminated with insecticide following house-spraying.

One of the most dramatic examples of pesticide accumulation along a food chain (bio-accumulation) was seen at Clear Lake, California. Starting in 1949, organochlorine pesticide (DDD) was added after a very careful calculation of the lake volume was made to ensure that no more than one-fiftieth of a part per million would be present. The purpose was to control the larvae of the gnat, *Chaoborus astictopus*, of which, although incapable of biting, the vast numbers of emerging adults were interfering with the use of the lake for recreation purposes. In 1954 and 1957 large numbers of western grebes (*Aechmophorus occidentalis*) were killed, and post-mortem analyses showed that they had accumulated DDD to the high level of 1600 ppm. Fish upon which the grebes fed were also found to have high levels of DDD. The pesticide had accumulated in plankton, then further by fish, so that the food of the fish-eating grebes was uniformly loaded with the organochlorine pesticide. In fact the initial DDD concentration in the water of 0.02 ppm was magnified 80 000 times in fish and grebes.

Blanket-spraying of wild vegetation to control insect vectors of human and livestock diseases has sometimes led to severe environmental damage, e.g. killing birds, reptiles and mammals including hippopotamuses; also many non-target insects. It must be pointed out, however, that although this is regrettable, such areas are usually recolonized by the wildlife, whereas if land freed of tsetse flies is farmed this leads to permanent habitat loss for wildlife.

It is possible to avoid such problems in tsetse fly control by restricting insecticidal spraying to the bottom 1.5 m of the tree trunks in the dry season and the bottom 3.5 m in the wet, because tsetse flies rest mainly on the lower sections of trees. Also less persistent compounds such as pyrethroids, applied at ultra-low-volume from ground or air, can be substituted for organochlorines, though about five applications at ten-day intervals are made necessary because fly puparia, which are unaffected by spraying, remain in the soil for as long as 4–5 weeks.

The human race is peculiarly devoted to its 'feathered' and 'furry' friends, and effects of pesticide on birds have no difficulty in reaching the headlines. However, there are also many less-publicized effects on lower animals which represent equally if not more devastating repercussions of the broad-spectrum toxicity of most pesticides.

One of these is the so-called 'resurgence' problem. A broad-spectrum chemical may be very effective against a highly mobile pest, but at the same time be equally or even more effective against other organisms, particularly natural enemies of the same or another pest. The sensitivity to pesticides of many of these beneficial insects is considerably higher than that of the pests, since they have not had to evolve the same armoury of enzymes to deal with natural toxins as have herbivores. The mobile pest re-invading the crop is then able to multiply without restraint from its natural enemies, and a far worse pest problem may result than was present before the pesticide was applied. Already quite a long time ago, Ripper (1956) was able to cite many examples of the resurgence problem, including the classic one of para-oxon and cabbage aphids (see Chapter 2). In Kenya, when rice fields were sprayed against rice stem borers (*Chilo suppressalis*), not mosquitoes, there was nevertheless a drastic kill of mosquito larvae. The rice fields were rapidly recolonized and the mosquito population increased and went beyond the pre-spray levels. This was because many larval invertebrate predators were also killed, but their populations took longer to recover and this allowed mosquito numbers to increase dramatically. This interesting phenomenon has been observed in other countries, including Japan. Such spectacular resurgences are often short-term and easily reversible, but there are other examples of slower, less spectacular but usually much more permanent phenomena. The appearance of new pests as the result of the use of pesticides is usually a long-term effect. The classic example of such a man-made pest problem is probably that of the fruit tree red spider mite (*Panonychus ulmi*), which was promoted from insignificance to major pest status in the 1940s. The problem resulted from the use of DDT against codling moth (*Cydia pomonella*) and the use of tar oils against overwintering eggs on apples. DDT killed the red spider mite's predators as well as stimulating its fecundity. Tar oil killed the eggs of predators and competitors, whereas winter red spider eggs selectively survived because they carry their breathing pores on an elongated tube rather like a snorkel which penetrated through the oil film.

Analogous is the creation of about 100 scale insect pests, referred to in Chapter 1, by the use of pesticide to control one such problem in Californian citrus orchards.

There are at least two non-target organisms in the crop other than natural enemies. First, bees are often important pollinators and are at considerable risk from insecticides. Although nearly all farmers and growers either avoid spraying when their own crops are in flower or warn beekeepers to close their hives, bees may suffer from insecticide drift, which can carry insecticide on to fodder crops and wild plants on waste land or hedgerows near the crop. Second, there is the effect of pesticides on the crops they are designed to protect. Sometimes the symptoms of this phytotoxicity are spectacular, as with the sudden leaf drop of 'sulphur shy' currant bushes when a sulphur-containing compound is used. The chemical manufacturer during the pesticide's development usually spots such dramatic phytotoxicity, and the chemical will then be diverted to the herbicide section. Some other expressions of phytotoxicity are much less obvious, and may only be discovered after the chemical has been in use for many years. Many pesticides cause some check to the growth of the plant and result in small reductions in yield, but farmers are rarely in the position to compare the yield of sprayed versus unsprayed crops. Other examples include an impaired crop flavour (taint), a reduction in fruit set and a tendency to accentuate the biennial cropping problem in apples. Phytotoxicity may also result from formulation. For example, addition of higher concentrations of wetter to wet the waxy surface of many fruits has sometimes caused local damage where the liquid has run off and left a drop, gradually concentrated by evaporation, at the base of the fruit (Fig. 5.2). It must be stressed that many of the long-term problems of effects on non-target organisms, particularly the larger 'wildlife', have now been reversed since the highly persistent organochlorines have mostly been replaced by much shorter-lived insecticides.

Nature fights back

To some extent greater knowledge and better testing have enabled industry to produce modern pesticides that are safer for humans and less damaging to the environment. The problem with pesticides that has yet defied solution is that

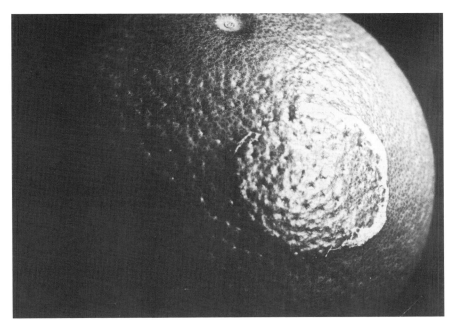

Fig. 5.2. Underside of an orange showing the scorch ring where insecticide has run down to form a large drop which has then slowly evaporated and concentrated the toxin at this point (The late F. Baranyovits).

pesticides are likely to lose their effectiveness after prolonged use, sometimes even after only a few seasons of use due to the development of insecticide resistance. How does this arise? Well, pest populations invariably have a genetic pool of widely differing susceptibility to the poison, and the use of pesticides creates a selection pressure on the population whereby the less susceptible individuals are left to breed the next generation (Fig. 5.3). These individuals have properties such as less permeable cuticles, faster storage of toxin in fat or a better equipment of enzyme systems for metabolizing the toxin. This is the commonest type of resistance and is often called physiological resistance. The properties conferring resistance are genetically inherited and can be passed on to their offspring. Originally rare in the population, the possession of these properties therefore becomes increasingly common, as shown in Fig. 5.3. The appearance of individuals apparently far more resistant to the pesticide than were present in earlier generations is due to the fact that they are becoming more common and are therefore likely to be included in a sample. Another type of resistance is termed behavioural resistance. This involves changes to the behaviour of a species, such as making it avoid contact with an insecticide

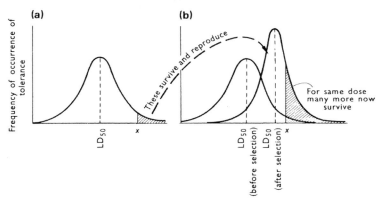

Fig. 5.3. Unnatural resistance and the development of resistance to insecticides. (a) Distribution of tolerance before selection. (b) Distribution of tolerance after selection compared with distribution before selection. *x*, the dose exerting selection pressure, kills fewer organisms in the population of progeny than in the population of parents. LD_{50} = dose lethal to 50% of the population.

due to its irritant effect, or avoidance due to non-contact repellence caused by a fumigant effect.

Certain groups of arthropods are notoriously prone to developing resistance. This primarily reflects the insecticide pressure we have put on them. As a proportion of the recorded cases of resistance across all insects and mites, the Lepidoptera, Coleoptera and Hemiptera (with of course many plant and animal disease vectors) each represent a little under a quarter. Mites/ticks and Diptera (including the most important human and animal disease vectors) each represent about one-eighth.

Speed of development of resistance depends on several factors, including the proportion of resistant individuals originally present, degree of isolation of the population from neighbouring populations not exposed to insecticide and the reproductive rate and generation time of the pest.

Resistance may even appear outside the targeted area. Drift or run-off of insecticide aimed at crop pests may be sufficient, when the toxin enters the water system, to cause mosquitoes to develop resistance due to larval contact with insecticidal waters. This has been reported several times.

In 1944, only 44 arthropod species were known to have developed resistance, but by 1990 more than 450 species were known to have become resistant to one or more pesticides. It is important to be able to detect resistance, especially in its early stages. The obvious approach is to dose samples of the population to be tested with increasing doses of pesticide, and compare the rise in mortality with increasing dose with what happens in a population known to be susceptible. This is normally done, either by dosing individuals individually with a drop of insecticide or, especially with smaller arthropods such as mites, mosquitoes or house flies, by allowing the insects a defined time of contact (usually an hour) with a treated surface or a range of surfaces with increasing concentrations of insecticide, followed by a holding period (usually 24 hours). The dose which kills 50% of the insects (known as the LD_{50}) is the datum calculated because it is the point on the dosage-mortality line which has the smallest error in estimation; growers would obviously aim for a much higher level of kill. The LD_{50} is therefore only a laboratory statistic; it is affected by length of contact with the poison, the interval (holding period) before mortality is measured, the temperature during that interval and many other things. Thus it is only useful for relative comparisons, i.e. between a known susceptible and a suspected resistant population, and to monitor changes in the susceptibility of a population over time.

One problem with this technique, which was how it was done for many years, is that with many insects it is surprisingly hard to find a criterion of 'death' at short intervals after dosing, and that variability in replicates increases after 24 hours. Also, for reasons stated earlier and others, mortality does not always show a consistent rise with increase in concentration. Furthermore, resistance tends not to be detectable until there is some resistance in about 10% of the population and, if something is to be done about it, this detection threshold is far too high. The biochemical mechanisms of resistance are now known for most groups of pesticides, and a more sensitive assay for resistance than measuring mortality is to detect the presence of the resistance mechanism. Resistance is most often due to the presence of enhanced levels of a detoxifying enzyme or of a target site insensitive to the pesticide. There are now well-established techniques for evaluating such resistance in individual insects, even ones as small as aphids. With automated handling of microtitre well plates, the availability of appropriate monoclonal antibodies and sophisticated methods for measuring enzyme kinetics, we are probably getting close to identifying resistance at a gene frequency as low as 0.01. However, resistance to an insecticide may not be caused by just one enzymatic mechanism. Hardness of cuticle and pesticide excretion are two mechanisms which may be operating at the same time but would not be detectable by such enzyme assays, though their contribution would be detected by the LD_{50} bioassay referred to earlier.

Another factor complicating the monitoring of resistance is that an insect may show resistance to more than one chemical. Often this is cross-resistance, where one resistance mechanism confers protection to more than one insecticide. However, there are many cases of multiple resistance, where the insect has independently accumulated several mechanisms of resistance to different chemical groups of insecticides.

Three questions that perhaps arise in the reader's mind are: 'Do not natural enemies similarly become resistant?' 'Why cannot we apply a dose of pesticide high enough to leave no resistant survivors to breed the next generation?' and 'Can't we stop the development of resistance?'

The answer to the first question is 'Yes, but much more slowly than pest species'. Predators and parasitoids tend to have fewer generations per year than pests, and therefore fewer opportunities for genetic selection. Also, the pesticide is applied by techniques specifically designed to make effective contact with the pest species. Natural enemies are behaviourally different and may often consume much untreated prey; they are also often more mobile and can therefore avoid pesticide better than the pest species can. Moreover, the numbers of natural enemies depend on how abundant the prey is, and therefore the prey has to develop resistance first before a change in abundance of the predator is noticeable. It is therefore not surprising that the first resistant predators found in the field were predatory mites preying on pesticide-resistant mite pests. The behaviour and spatial occurrence of the predatory mites and their prey are also fairly similar. Although the apparent survival of such mites following pesticide application was noticed as early as 1953, it was not until 1970 that the resistance was confirmed in the USA. Experiments have been carried out since 1949 to try to select for insecticide resistance in natural enemies under laboratory conditions, with a view to releasing such resistant strains into crops to be treated with the relevant insecticides. This idea of inducing insecticide resistance in predators and parasites has recently gained considerable momentum. The appearance of pesticide-resistant predatory mites in the field has been followed by further reports of other species. There are at least seven reports of resistant natural enemies in the field, of which five are mites (three *Typhlodromus* and two *Amblyseius* species), one a ladybird (*Coleomegilla maculata*), one a parasitic fly (*Ophyra leucostoma*, Anthomyiidae) and two parasitoid Hymenoptera (*Macrocentrus ancylivorus* and *Bracon mellitor*, both Braconidae). Another idea, albeit controversial, is that with genetic control methods such as the release of sterile males to mate with native females (see Chapter 9), if the released sterile males are made insecticide resistant this will allow them to survive any spraying programme in an integrated control approach. Today, genetic engineering has the potential for us to visualize transferring genes for resistance to insecticides from pests to natural enemies (see Chapter 7).

The answer to the second question is that there are already remarkably resistant individuals (although they are rare in the natural population) before they encounter any insecticide. Although an excessive dose would probably kill even these individuals, the crop plants themselves usually have some limit of tolerance to an insecticide before damage to the plant occurs. There is also a limit as to how much pesticide can be brought in contact with an insect, set by the latter's size.

As regards the third question, although it may be possible to delay the appearance of resistance, it is probably only possible to delay it for ever in that very small proportion of cases where resistance can be developed to two chemicals independently, but such resistances are negatively correlated. There is a well-known example in brown planthopper of rice (*Nilapavarta lugens*), which shows such negative cross-resistance to two analogues of the carbamate insecticide carbaryl. Another example occurs with black flies (Simuliidae), where chlorpyrifos/chlorphoxim (both organophosphate insecticides) resistance has a low level of negative cross-resistance to permethrin. This phenomenon has been exploited by rotating these insecticides in the Onchocerciasis Control Programme in West Africa, and this rotation has worked successfully for many years. There is also anecdotal evidence for other negative cross-resistances between organophosphate and pyrethroid insecticides. This has led to the suggestion that, where mosquitoes have developed resistance to pyrethroids, then organophosphates should replace the currently used pyrethroids for impregnating bed nets. Otherwise, there are a number of strategies which have been proposed for at least delaying resistance.

Level of kill

Different scientists argue for a low or for a high kill strategy. The former is based on models suggesting that a high kill strategy will inevitably lead to resistance and will accelerate this by denying the possibility of resistance being diluted by mating with more susceptible individuals. The proponents of a high kill strategy argue that it is dangerous not to kill the individuals with partial resistance in the heterozygous state. That there is still argument about this simple alternative of strategies is worrying!

Order of using different compounds

There is a similar lack of consensus about another simple choice – whether several compounds will in total last longer if used in rotation or sequentially (i.e. not used until the previous one fails). Here, in spite of the disagreement about the theory, field practice suggests that the compounds will usually fail earlier through multiple resistance if used in rotation than sequentially, though the pest mites seem the exception to this rule.

Switching the targeted life stage

This strategy is only relevant for the higher (endopterygote) insects, which have morphologically very different larval and adult stages as well as a resting pupal stage. Insecticides are very often used against them either in the larval or the adult stage. Since the larval and adult stages usually have totally different food requirements, it is likely that their arsenal of detoxifying enzymes will also differ. Switching the life stage targeted once resistance in the other stage appears should at least prolong the useful life of the compound.

Spray window restrictions

Farmers have often switched to a new insecticide when their previous one shows resistance at the end of a season, and have then begun with the new insecticide the next year. This may be a grave mistake hastening resistance to the new insecticide. Particularly where the pest is migratory or has other hosts which do not receive insecticide, and thus may have encountered susceptible mates, the pests appearing at the start of a new season may not have the insecticide resistance of the pests at the end of the previous season. The lesson is to restrict the use of the new insecticide each season until it is shown to be necessary. Restricted spray windows of between 6 and 14 weeks for the very effective synthetic pyrethroids have been used on cotton in several different countries to ensure that only one of the annual generations of bollworm is exposed to selection for resistance to these insecticides. Experience, however, tends to show that the delay in resistance development achieved by this strategy is not as long as might have been hoped.

Pesticide rejuvenation

If the mechanism of resistance is understood, it may be possible to add another compound (a synergist) which blocks that mechanism biochemically to the formulation. This has been a solution keenly sought by industry. Thus rejuvenation of synthetic pyrethroids, to which resistance has developed, has been attempted by adding the synergist piperonyl butoxide. At first sight it seems a great idea! However, when resistance develops to this synergized pesticide, one has a real 'super-bug' capable of detoxifying huge amounts of pesticide and probably a large range of related compounds. It is perhaps not such a good idea, after all.

Diversifying selection pressures

The selection pressure of insecticide on a pest population will be greatly reduced if there is more than one source of mortality. This is the role of Integrated Pest Management, which we describe later in Chapter 13, and is of course one way of using the 'low kill' strategy with a much reduced danger of heterozygotes surviving.

The resistance race

The 'tolerant pest strain' problem is so serious because there is a real danger that the appearance of such strains will outstrip the production of effective pesticides. As the problem seems to be the inevitable consequence of pesticide use, it is not surprising that all groups of pesticides are affected. Over half the world's pests show a tolerance to at least one major group of insecticides. The production of new chemicals has never been a rapid process; moreover, nearly all the pesticides we use share some similarity, usually in their mode of attacking the biochemistry of the nervous system. The economics of pesticide use mean that there is a ceiling to the price which farmers or public health authorities can pay, and this gives a ceiling to the profitability a new pesticide can achieve before it is forced out of use by the appearance of resistance to it. The costs of testing and developing a new pesticide soar with inflation and increased demand for safety testing. As pointed out on p. 57, the current costs must be approaching US$ 120 million, and it is not unusual for 6 of the 20 years of patent life to expire before the product is even marketed (Fig. 3.2). Companies which do not carry the heavy overheads of development can then manufacture chemicals whose patents have expired, and this will cause a drop in price of the product. The chances of retrieving the development costs with an adequate profit in the short residual patent life and before resistance develops are daily becoming slimmer, and it is in no way surprising that several companies have 'opted out' of the development of new products.

The other road

This chapter comes at the end of Rachel Carson's book, where she urges us to abandon pesticides and seek 'The Other Road', particularly via biological control of pests. The succeeding chapters explore this other road, particularly to see how far a general alternative to chemical control emerges. Rachel Carson pondered what man's fate would be when he came to the end of the insecticide road. By the time her book was published, man had already found the insecticide road to have run out in at least three crop situations. An account of those events and how they were solved will be found in Chapter 13.

6

Environmental/cultural control

Introduction

Mankind had to try and manage insect problems for centuries before insecticides became available, biological control was purposefully introduced or the many other approaches described in this book were developed. The mainstay of these earlier efforts was to make the environment unfavourable for the development of pests. For human and animal disease vectors, the environment was usually the wild vegetation around habitations as well as the habitations themselves, and the term 'Environmental control' is generally used in these contexts.

The same phrase is, however, virtually unknown to agricultural entomologists. For them, the environment which is to be made less favourable for the pest is, more often than not, a crop, and the techniques of environmental modification are manipulations of the various cultural practices (ploughing, planting, harvesting etc.) carried out by farmers. Pest control based on these practices is always called 'Cultural control'; therefore we have used the two words 'Environmental' and 'Cultural' in the heading to this chapter. It should be pointed out that there is yet another aspect of 'Environmental control', where the microclimate in closed systems (e.g. grain-storage containers) is made lethal to pest organisms. This aspect is dealt with in Chapter 12 among 'Physical methods'.

History of environmental/cultural control

As pointed out in the Introduction, these controls were mankind's chief weapon against pests before the arrival of the powerful modern synthetic pesticides, and thus represent most pest control used in the long history of our war against insects. As there seems general agreement that we can put a date of 10 000 years ago to the start of agriculture, we can illustrate how long mankind relied on environmental/cultural controls by condensing this 10 000 years down to just one calendar year. In this time frame, cultural control ruled till 6 a.m. on 29 December. At this point (the early 1940s in real time) the *Review of Applied Entomology*, an abstracting journal published since 1913, shows a sudden switch from experiments on cultural control to insecticide trials. Only 14 hours later (on our condensed time scale) the overuse of the 'new' method (pesticides) had stimulated the publication of *Silent Spring* (see Chapter 5). Of course, such calculations, based on the history of agriculture, underestimate the proportion of the insect war fought without insecticides; as pointed out in Chapter 1, insects bit mankind several thousand years before the dawn of agriculture.

The reason for the sudden acceptance of insecticides in both medical and agricultural scenarios was, of course, that the new insecticides offered a previously unattainable level of control. Given that environmental and cultural measures were often in conflict with other needs of mankind, also that they were often labour-intensive and thus expensive in the developed world, it is not surprising that they were often abandoned or given a much lower priority once insecticides became easily available.

Conversely, until recently it has been labour and not pesticides that have been cheap in developing countries in the tropics. Thus environmental/cultural measures for pest control are still widely practised there.

Pesticides originated for the developed world in the developed world, much of which is temperate in climate. Thus pests and vectors have fewer generations compared with the tropics. In humid areas, or where irrigation is used to maintain ornamental and crop vegetation in the dry season, the long quiescent season characteristic for pests in temperate climates (winter) is missing.

In reducing the emphasis on environmental/cultural controls in the tropics, mankind has ignored the important role these controls have of greatly reducing average pest densities in a region, even though they usually give inferior control to pesticides. So both climate and the relaxation of environmental controls contribute to the dramatically more rapid growth of insect populations in tropical than in temperate climates. The problems of insects resistant to pesticides in many tropical areas have already shown that they cannot be restrained where environmental/cultural controls have been replaced by total reliance on pesticides. An important principle of Integrated Pest Management (IPM) for the tropics is therefore the retention of environmental/cultural control measures.

A second principle of IPM relevant to such controls is not related to existing practices, but to the IPM consequences of new ones. Modifying the environment carries the potential for 'problem trading'. Control aimed against one problem may well improve conditions for another. Such problems that environmental or cultural measures can create often become evident, not only in purposeful attempts to reduce insect problems, but equally in changes to the pest spectrum that may occur when environmental or crop management practices change for other reasons. Both sides of the 'coin' are reviewed in the account that follows.

This chapter reviews environmental and cultural controls aimed directly at reducing the numbers of pests and disease vectors, although some such measures may also affect natural enemies for better or for worse. Environmental modification aimed directly at promoting biological control is the topic of 'Conservation Biological Control', and we have dealt with this in Chapter 7 as a form of biological control.

Sources of environmental/cultural control

Mechanical disturbance of the environment

This usually involves heavy machinery to either eliminate habitats or make them unfavourable for noxious insects.

Many insects live or hibernate in suitable temperature and humidity conditions relatively near the soil surface. These conditions can be disturbed by ploughing, which creates temporary drought conditions in the upper soil layers and may even expose larvae and pupae to the full radiation of the sun. Birds will eat many of these insects (Fig. 6.1), and pigs have even been brought on to ploughed fields for the special purpose of picking out and eating white-grubs (scarabaeid beetle larvae, pests of cereals). Other pupae and eggs may be buried by ploughing to a depth from which they fail to reach the soil surface after emerging. Then other individuals will be killed mechanically by rough contact with soil clumps; and root aphids (e.g. cereal root aphids) will suffer from the break-up of the ant colonies which tend them. Little is known about the effects of ploughing on beneficial insects. However, it is known that adult cereal leaf beetles (*Oulema melanopa*) disperse from the fields to overwinter, whereas their larval and pupal parasites remain in the soil and ploughing in the spring destroys many.

Compacting the soil with a heavy roller is a cultural measure for limiting the between-plant movement of some larger soil insects such as beetle larvae.

Fig. 6.1. Birds following the plough to predate insects brought to the surface. North-east Lincolnshire, UK (H. F. van Emden).

Distribution and abundance of tsetse flies is determined largely by types of vegetation and microclimate. Their dependence on such requirements formed the basis of control through clearing away vegetation either partially or completely, especially along rivers and lake shores where the so-called riverine tsetse flies rest. Complete, or ruthless, clearing resulted in scrub or woodland being replaced with grassland habitats unsuitable for tsetse flies. The work was costly and labour intensive. To reduce costs partial or discriminate clearing was introduced. This relied on greater knowledge of the exact resting sites of the flies and as a consequence only the understorey was removed, and most of the taller trees were left. In the past such destruction of vegetation has achieved considerable success in controlling tsetse flies. However, destruction of vegetation is no longer ecologically acceptable. Nevertheless, clearing vegetation for human settlements or felling trees for firewood, or even destruction of vegetation by dense populations of elephants can have the same effect, that is reducing fly populations.

A simple approach to eliminating sites for mosquito larvae is to fill in their habitats, ranging from water-filled tree holes to ponds and small marshes, with rubble, earth or sand to completely eradicate breeding sites. Larval habitats such as ponds and marshes can be drained. An important advantage of filling in, draining or eliminating breeding places is that this can result in permanent control, but this approach is not always feasible. Some important pest

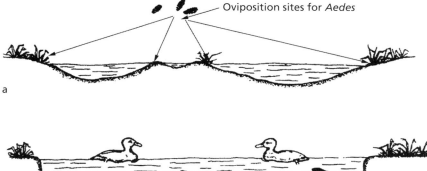

Oviposition sites for *Aedes*

a

b

Fig. 6.2. Diagrammatic representation of (a) shallow pond or marshy area with exposed wet soils suitable for oviposition by *Aedes* mosquitoes; (b) the same habitat excavated to leave deeper water and vertical banks unsuitable as *Aedes* oviposition sites. Stocking with fish and aquatic birds will also reduce mosquito breeding (M. W. Service).

mosquitoes and vectors breed in tree holes, but these may be high up in the trees and are therefore difficult to locate, or they may be too numerous for this method to be practical. It is impossible to fill in all the scattered ephemeral pools and puddles that appear in the rainy seasons, and larger water collections such as ponds may provide essential water for domestic purposes or watering places for livestock.

A further mechanical intervention is to modify breeding places to become unsuitable for mosquitoes, rather than eradicating the habitats. Instead of draining marshy areas which support numerous small pools, many of which being transient fail to support predators but are colonized by mosquitoes, they can be excavated to form areas of deep water with well-defined vertical banks (Fig. 6.2). Such impoundments can be relatively small or quite large and become important water reservoirs. Creation of such large expanses of water is usually inimical to mosquito breeding. Ducks and other wildfowl may be attracted to such sites and fish introduced. Some of the more attractive habitats can be used for recreational purposes. But it must be realized that one type of habitat, a marshy area, has been eliminated and replaced by a large pond. Such habitat changes may not be ecologically acceptable. As another example, mosquitoes breeding in isolated small pools and marshy areas formed at the edges of small meandering streams may be eliminated by realigning water courses to increase water flow and prevent the accumulation of pockets of static waters. Removal of floating or rooted aquatic vegetation will prevent breeding by *Mansonia* species of mosquitoes, whose larvae and pupae need to attach themselves to vegetation in order to breathe.

The potential for undesirable side-effects of environmental/cultural controls is shown by several examples of the mechanical measures just mentioned. The removal of aquatic vegetation to control *Mansonia* may make the habitat attractive to other species such as anopheline mosquitoes, leading to replacement of one vector species by another.

Minimum tillage (whereby ploughing is given up in favour of the use of herbicides to kill any weeds on the soil surface followed by the direct drilling of cereals into slots cut in the otherwise undisturbed soil surface) was a major development in cereal crop management in the 1970s. Where it was introduced, it created numerous pest problems. Some of these relate to the earlier drilling of cereals permitted by this method, and will be discussed later in this chapter. However, slugs and cutworms (larvae of noctuid moths) benefited particularly from the lack of soil disturbance in minimum tillage systems, the mat of (albeit dying) weeds on the field surface providing suitable conditions in addition to the lack of disturbance of the soil. Another problem with minimum tillage has been that of the stem-boring frit fly (*Oscinella frit*). This insect can develop large populations in grasses and the larvae migrate into winter wheat seedlings when wheat is drilled into a herbicide-treated sward in the common rotation of cereals after grass. This problem was never serious when the old sward was ploughed in late summer and left fallow until the sowing of spring wheat the following year.

Irrigation

Irrigation is a common practice in many crops, and it can be manipulated for pest control purposes. Small pests such as aphids are easily washed off plants by overhead irrigation, and the pressure of swelling soil particles in saturated soils may kill soil insects. Additionally, ample water availability causes physiological changes in plants; some sucking insects such as aphids and thrips tend to do badly on well-irrigated plants and benefit from periodic wilting of the plants. Raising the water level in rice paddies had been used to suffocate eggs of armyworm (*Spodoptera* species) and to drown larvae of the moth rice borer *Scirpophaga incertulas*. Active insects such as shield bugs (Pentatomidae) may be driven off the rice into the water, and subsistence farmers in Asia have been known to then put ducks onto the paddy to eat the swimming pests.

Fig. 6.3. Bases of rice plants showing the dry nature of rice fields when water has been drawn off as part of a programme of intermittent irrigation (M. W. Service).

When rice paddies are irrigated (see Fig. 1.3), whether in the tropical Americas, USA, Europe, Africa or Asia, this provides prolific breeding places for many species of mosquitoes, including *Anopheles* species that are malaria vectors. Indeed, in Portugal in the 1930s the term 'rice malaria' was coined for malaria associated with mosquitoes breeding in rice fields. In Asia *Culex tritaeniorhynchus*, a vector of Japanese encephalitis (JE) which infects humans and pigs, is very common in rice fields. There are more than 40 arboviral diseases, such as JE, spread by mosquitoes breeding in rice fields; admittedly many are rather unimportant.

Considerable attention is being focused in some parts of the world, e.g. China, on the 'alternate wet and dry cultivation of rice' (intermittent irrigation), to reduce mosquito breeding. The procedure involves water remaining on fields for less than a week, an insufficient time for mosquito larvae to complete their aquatic development. In Hunan province (China), about two weeks following plantation of rice seedlings, the paddy fields are dried out (Fig. 6.3) for 24–48 hours; this is then repeated every three to five days. Commonly there may be 21–6 floodings during a 100-day rice growing cycle. This has apparently resulted in 81–91% reductions in the numbers of immature stages of *Culex tritaeniorhynchus* and an 84–6% decrease in numbers of

Anopheles sinensis, an important malaria vector. Reductions in vector populations have been accompanied by decreases in transmission of malaria and JE. However, this approach does not always seem to work. For instance in Japan, despite intermittent irrigation causing marked reductions in *C. tritaeniorhynchus* larvae, this has not always been accompanied by any apparent decrease in the numbers of adult mosquitoes biting pigs, which are important amplifying hosts for JE. In China, intermittent irrigation is widely accepted by farmers because they believe it increases crop yields (by about 14%); there is in reality little evidence for this.

If soils become waterlogged intermittent irrigation is unlikely to have any effect on reducing mosquito populations because after draining off the water isolated pools will remain and support mosquito breeding.

In China the introduction of fish, such as carp (*Cyprinus carpio*), into rice fields has also helped to reduce mosquito breeding and in certain areas also the incidence of malaria. Stocking rice fields with fish is popular, not because it reduces mosquito numbers, but because fish faeces help fertilize the rice crop. So the farmer, in addition to harvesting about 500 kg of rice from his typical 15-hectare holding, gets about 50 kg of fish which he can sell at a considerable profit.

It should be appreciated that farmers keep fish and use intermittent irrigation because of the financial benefits they bring, not because these practices may reduce disease transmission.

Where irrigation is essential to growing a crop, as in naturally arid places in California and large areas of the American Middle West, the only lush vegetation in the region (i.e. the crop) may act as a magnet for pests. It is a well-known phenomenon in dry areas that pest incidence on cotton rises dramatically following irrigation. In California and Peru, however, irrigation has also enabled a pest/natural enemy complex to persist and reduce the importance of bollworms (*Heliothis* and *Helicoverpa* species) as pests.

The most bizarre attempt to control pests with irrigation occurred in the late nineteenth century when vine growers in France were threatened by the dreaded phylloxera (*Daktulosphaira vitifoliae*, a close relative of aphids), which had been accidentally introduced from North America. They took the desperate step of flooding their entire vineyard for long periods to suffocate the pest, in spite of the damage such flooding did to both the plants and the soil.

Fertilizer and other soil ameliorations

The belief that vigorous plants are less attacked by pests is one of the foundation stones of so-called 'organic' farming, and it is far from being an erroneous concept. Rapid, healthy plant growth can reduce pest damage in four ways:

- Rapid growth shortens any susceptible plant growth stage. It therefore induces resistance against pests such as stem borers, to which seedlings have a relatively short window of susceptibility before the tissues harden.
- It may lead to the maximum expression of some chemical resistance factors.
- It will allow maximum compensation for damage by the plant. For example, good root systems would clearly withstand root grazing by pests where weak root systems would not. Another example concerns the shot-hole borer (the beetle *Xyleborus fornicatus*) on tea in Sri Lanka, where fertilizing the bushes with nitrogen successfully reduced damage. The stimulation in growth enabled the bush to form new tissue as a support bracket over the beetle gallery so that breaking of the branches at the gallery as tea pluckers passed through the plantation no longer occurred.
- It can promote uniformity and density of the crop stand. This can discourage pests such as the chinch bug (*Blissus leucopterus*), which is most abundant where the crop stand is somewhat thin. Similarly, aphids occur in smaller numbers where the crop is denser; this is because fewer winged immigrants land where less bare ground is exposed.

However, just as fertilizer produces a more nutritious plant for man, so many insects may also benefit. Aphids, leafhoppers, mites, thrips and leaf-mining grubs have all been found to breed or develop more rapidly on plants given good nitrogen fertilization. By contrast, there is some evidence that manuring with potassium and phosphate may reduce the incidence of some pests, and with aphids, which are sap feeders, good potassium fertilization can reduce nitrogen available in the sap without impairing the value of the leaf protein. Here is a field ripe for useful and relatively simple experimentation – how far can we 'induce' plant resistance (see Chapter 11) by physiological treatments to plants?

Mulching (Fig. 6.4), i.e. covering the soil, usually with plant debris, brings many agronomic benefits, especially the retention of higher humidity in or near the soil and the suppression of weeds. In coffee, thrips are rarely a problem in the more humid conditions of mulched plantations; just one season without mulch may elevate this insect to pest status. Damp conditions created by mulching may also be favourable to insect parasitoids; thus mulching in coffee increases the biological control of the Antestia bugs (*Antestiopsis* species). Still with coffee and Antestia bugs, pruning management is a further weapon. This bug does less well where humidity in the canopy is reduced by pruning. Lower humidity, unfortunately, also makes the environment less suitable for parasitoids of the pest, but this can be minimized by careful

Fig. 6.4. Banana plantation mulched with dead banana leaves, Uganda (courtesy of S. R. Gowen).

timing of the pruning operation and by leaving the prunings on the ground as a mulch.

Sanitary measures

Sanitary measures are perhaps the most important element of environmental control of human and animal disease vectors.

Many man-made larval habitats such as abandoned tin cans, metal drums, disused water-storage pots and old vehicle tyres (Fig. 6.5) can be removed, an approach that is sometimes referred to as source reduction. Water-storage containers (see Fig. 2.7) cannot be discarded because they provide essential domestic water, although in theory they can be covered with plastic mosquito screening to exclude ovipositing mosquitoes, such as *Aedes aegypti*, and yet still allow them to be filled with water. However, in practice it is notoriously difficult to get communities to cover such potential breeding places (see Chapter 12). The introduction of a piped water supply should help reduce people's dependence on water pots, but this is frequently not the case because it may be unreliable, so water pots often continue to be used as a type of

Fig. 6.5. Motor-vehicle tyres in an urban tyre dump in the USA. When filled with rain water they become breeding places for mosquitoes such as *Aedes aegypti* and *Aedes albopictus*, vectors of dengue fever (M. W. Service).

insurance policy. It also takes a lot to change people's behaviour and persuade them to abandon the tradition of keeping water pots.

In many parts of Asia coconut husks are soaked in coir pits (Fig. 6.6) prior to the fibre being used in the manufacture of coconut matting. These pits become infested with larvae of the mosquito *Culex quinquefasciatus*, an important vector of bancroftian filariasis. Understandably the pits are situated very near to the owners' houses, consequently people are exposed to large numbers of hungry female mosquitoes and stand a good chance of becoming infected. People are loathe to abandon these convenient coir pits and dig new ones further afield, but we know that if this was done it would lessen transmission risks.

Environmental sanitation is also important in house fly control. Unsightly and unhygienic rubbish dumps, usually containing food and other decomposing organic matter, are commonly found in villages as well as in urban areas. They provide ideal breeding places for house flies and other synanthropic flies. Environmental sanitation aims at dramatically reducing house fly numbers by minimizing their larval habitats, such as accumulated rubbish. For example, domestic refuse should be placed in strong plastic bags with their openings

Fig. 6.6. Coir pit in Sri Lanka where coconut husks are soaked. This creates ideal larval habitats for the filariasis vector, *Culex quinquefasciatus*. Note the proximity of human habitation (M. W. Service).

securely tied, or in dustbins with tight-fitting lids. Collections of household refuse should be at regular intervals, in hot weather at least weekly. Rubbish also can be burnt or buried. If it is buried in pits the refuse should be covered daily with a 15-cm layer of soil, and when the pit is almost full covered with 60 cm of compacted soil. This is necessary to prevent rodents being attracted to buried decomposing rubbish and digging it up. This is the theory, but in practice rubbish continues to provide attractive, as least to flies, breeding places, and not surprisingly this often leads to fly-borne diarrhoeal infections, especially in children.

Sanitation is also important in managing the pests of crops, and effective destruction of crop residues which harbour pests (and incidentally also plant diseases) is perhaps the single most important cultural control measure in agriculture. Stalk destruction is commonly practised in maize against the stem borers *Busseola* and *Heliothis*; the latter, however, pupates in the soil and stalk destruction should therefore be combined with ploughing. Such destruction removes residual pest populations and eliminates plant debris on the soil surface in which many pests find shelter for hibernation. Examples of such pests are flea beetles (Alticinae) and whiteflies (Aleyrodidae) of brassicas.

Although crop residue destruction is particularly important in cereals, there is a serious sociological constraint to its implementation because peasant farming communities may rely very heavily on straw and other plant materials for fencing or roofing. However, this then creates resting sites for various medically important insects including (in Latin America) triatomine bugs, the vectors of Chagas disease. In fact the replacement of thatched roofs by corrugated metal or fibrocement roofs, the cementing over of mud floors and the plastering of walls to render a more or less smooth surface are important environmental strategies for reducing transmission of Chagas disease.

A classic case of the importance of sociological considerations occurs with deep-water rice, an amazing rice grown on the flood plain of Bangladesh and which can extend its internodes as fast as the floods rise to keep the ears above the water. The yellow-headed borer (*Diatraea centrella*), the caterpillars of which tunnel in the long stems below flood water level, could easily be controlled by burning the straw after the harvest of the crop once the floods have receded. However, with a virtual absence of trees on the flood plains, the straw becomes essential as fuel for cooking the very rice grains it once supported.

Two other crops where crop sanitation is used are cocoa and banana. In these crops respectively, stems are peeled to control rhinoceros beetle (*Oryctes*) or banana weevil (*Cosmopolites sordidus*).

Destruction of crop residues of cotton followed by a gap of several months before cotton is again planted is mandatory in many countries in the world, though not always enforced. It would be particularly effective against the pink bollworm *Pectinophora gossypiella*, since this pest does not have any wild hosts to maintain the species if the crop population is destroyed.

Another aspect of clean crop management is 'roguing' – the removal and destruction of infected growing plant material where there is danger of spread to other parts of the crop. Before the advent of adequate plant resistance to the pest, the control of reversion virus spread by the blackcurrant gall mite (*Cecidophyopsis ribis*) was largely dependent on the removal and burning of infested bushes. Roguing plants attacked by the sisal weevil (*Scyphophorus interstitialis*) is still a component of control of this pest in the tropics. Similarly, cutting out branches on which the larval frass is visible has controlled the yellow-headed cerambycid borer (*Dirphya nigricornis*) of coffee.

The several genera of fruit flies (family Tephritidae) and the coffee berry borer (the scolytid beetle *Hypothenemus hampei*) are pests which often emerge from fallen fruits on the plantation floor to infest new fruit. Removal of fallen fruits is therefore an effective, though labour-expensive, sanitation measure.

Reduction of alternative hosts

Vectors of human diseases often also feed on alternative hosts, and these are usually wild fauna. Thus, as deer support large populations of ticks (vectors of Lyme disease), attempts have sometimes been made to exclude them from parks and other recreational areas, for example by fencing, but this is not always appreciated by the public who wish to see deer in these places.

At one time control campaigns in Zimbabwe, Zambia and some East African countries involved killing, in selected areas, game animals that might provide a source of blood-meals for tsetse flies, or be reservoir hosts of trypanosomiasis. Clearly wide-scale and often indiscriminate slaughter of animals is no longer acceptable in a world increasingly sensitive to wildlife conservation.

Reminiscent of the killing of game animals for tsetse fly control are some environmental measures against phlebotomine sand flies. These are among the smallest biting insects, some are only 1.3 mm in length, but they can be annoying biters and transmit various forms of leishmaniasis, a collection of tropical diseases that also occur in warmer temperate areas such as the Mediterranean. Although leishmaniasis is mainly controlled by chemical methods directed against adult sand flies, because most transmission involves reservoir hosts such as rodents and dogs, in some areas control is focused on killing such animals. For example, in North Africa clearing vegetation around villages reduces gerbil populations, while in parts of Brazil, China and the Mediterranean area infected dogs and other reservoir hosts are destroyed.

In agriculture, undesirable alternative hosts are usually local weeds. These may be related to the crop, such as the malvaceous weeds often found near cotton fields. The destruction of such reservoirs for pest populations is often recommended, e.g. that nearby free-growing cotton and related weeds should be eliminated to reduce populations of the cotton-stainer bug (*Dysdercus fasciatus*) in cotton fields. The reservoir weeds may also be quite unrelated to the crop. Around cotton fields in Africa, the cotton aphid (*Aphis gossyppii*) feeds on over 20 weeds in many different familes. However, weeds outside the farmer's boundary are just as important as those within.

Diversionary hosts

Weeds and even sacrificial crop plants can be used as 'traps', to divert a pest from the main crop area, and which are intended to be especially heavily damaged by pests and often destroyed well before harvest. The land occupied, which in some instances may need to be in the order of 5–10% of the whole crop, inevitably involves some reduction in crop yield.

If insect pests can be concentrated in particular small areas of a field, they can then be destroyed with locally applied pesticide or some other technique to which insects are unlikely to develop resistance, such as ploughing in, feeding the vegetation to livestock or the use of a flame gun. Such concentrations of pests may be induced by position (e.g. edge rows for swede-midge *Contarinia nasturtii*), by planting taller plants at the edge of the crop to filter out flying insects, by earlier sowing (e.g. against the corn earworm *Helicoverpa zea*) and by spraying with attractants or choosing especially attractive plants as the trap crop (e.g. kale for certain bug pests of cabbages). The application of chemicals (often plant-derived) to attract or repel insects has led to the development of the 'push-pull' strategy, whereby a 'pull' from the sacrificial trap crop is coupled with a 'push' from the crop to be protected. We cannot resist adding here that the original scientific jargon for this approach was 'stimulo-deterrent diversionary strategy'!

A particularly ingenious example of a trap crop is the use in Canada some 40 years ago of a non-crop trap plant (brome grass *Bromus sterilis*) planted in a 15–20-m strip around wheat fields to control the stem-boring sawfly *Cephus cinctus*. The adults did not penetrate into the wheat crop but laid eggs on the brome grass in which many larvae developed per stem. It was not necessary or even advisable to destroy the grass, for the grubs cannibalized one another, and even most of the eventual survivors failed to survive to maturity in the grass, although their parasites were able to emerge. Thus the brome grass became a factory for converting pest biomass into natural enemies for the crop!

Western flower thrips (*Frankliniella occidentalis*) is a polyphagous pest of glasshouse crops which is very hard to control with insecticides. However, it is attracted into tubular flowers such as those of *Petunia*, and placing pots of these (Fig. 6.7) among the plants to be protected can usually keep the pest below damaging levels. It is normally recommended that the flowers be picked and burnt from time to time, but in our experience this is not really necessary.

It was mentioned earlier that wild and domestic animals can often act as alternative hosts for insect vectors of human diseases. This has led to a control approach called 'zooprophylaxis', defined by the WHO as involving 'the use of wild or domestic animals, which are not reservoir hosts of a given disease, to divert the blood-seeking mosquito vectors from the human host of that disease'. The concept is not new. As early as 1903 it was suggested that in northern Italy domesticated animals directly protected people from mosquito bites. Basically the idea is to keep livestock in areas adjacent to people so that the mosquitoes bite these animals, usually cattle, in preference to humans.

Fig. 6.7.   Potted *Petunia* plants among glasshouse plants to trap thrips (H. F. van Emden).

This strategy has been practised in several countries to control malaria. Some believe that increased numbers of livestock might have unintentionally helped the gradual decline in malaria in northern Europe and much of the USA.

There are, however, two sides to the coin. It may be true that increased numbers of farm animals attract malaria vectors that otherwise might have bitten people. On the other hand, the introduction of animals might increase vector population size because they provide a readily available additional blood-source for hungry mosquitoes, and as a consequence this could lead to increased biting on people and higher malaria transmission. There are other considerations, for example in the dry season watering holes may have to be dug for cattle and these will likely support mosquito breeding, also the numerous water-filled hoof-prints that are created are ideal breeding places for anopheline mosquitoes. Some people have advocated that regularly spraying cattle with insecticides such as deltamethrin or permethrin will increase the efficacy of zooprophylaxis by killing mosquitoes attempting to feed on them. But there is the danger that mosquitoes will be deterred from feeding on sprayed animals and be diverted to feeding on people.

There are several reports in the literature purporting to show that increased animal populations can reduce disease transmission, and other reports suggesting the corollary, that is reduced numbers of animals can lead to increased transmission. But there are also reports which appear to show that, after introducing livestock close to human habitations, biting and/or disease transmission has increased. There is usually insufficient good evidence to know what effect nearby livestock has had on the numbers of mosquitoes biting people. Because of the many variables and ecological complexities it is not easy to devise convincing experimental trials that will unequivocally show what happens. The jury is still out.

Multiple cropping in agriculture

Before intensive agriculture, farmers tended to grow several crops on one unit of land. Such multiple cropping is still common in peasant agriculture in the tropics. The practice arises partly from cultivation purely to meet family demand and partly from the break-up of larger crop areas by inheritance traditions. Such strip farming systems therefore tend to persist, even without any understanding of their valuable pest-reducing characteristics. Either the area is divided into relatively narrow strips of different crops, or low crops are grown either under or in between the rows of taller crops (intercropping) (Fig. 6.8). In intercropping, the low crop reduces weed competition by covering the ground rapidly, and prevents soil erosion and water loss. Both strip farming and intercropping often reduce pest attack. In strip farming, the intervening strips of a non-suitable food may prevent movement of pests from one strip of a crop to another or from one suitable crop to a different one. Moreover, where two crops harbour unspecialized natural enemies, these can move over on to a neighbouring strip if pests build up there. The abandonment of strip farming in Peru some 40 years ago has been cited as one reason for the bollworm outbreak on cotton there, and certainly re-diversifying the cotton agro-ecosystem there (Fig. 13.1) greatly reduced the incidence of the pest. The choice of adjacent crops is, of course, more important than the simple decision to diversify. Juxtaposition of wheat and maize, for example, would actually intensify the problems of shared pests such as chinch bugs (*Blissus* species) and eelworms (nematodes), whereas separating the crops by a strip planted with potatoes would reduce pest damage on the cereals. Strip cropping is of course unacceptable in highly mechanized farming; moreover, some pests (e.g. some grasshoppers) lay eggs at the edges of crops and can become a serious problem when, as with strip farming, the edge forms a large proportion of the crop.

Intercropping seems to have three main effects on insects which result in lower pest numbers:

- It may reduce pest damage by attracting the pests to a less valuable crop, or one where the pest is less serious for some reason. One example is the

Fig. 6.8. An intercrop of legumes between maize in Tanzania (H. F. van Emden).

mixing of maize and cotton to achieve control of the shared lepidopteran pests *Helicoverpa armigera* and *H. zea*. This tactic can backfire if not timed perfectly; the formation of young bolls on the cotton must coincide with the tasselling of the maize. *Helicoverpa* is attracted to the maize tassels, but is a much less serious pest on maize than cotton, because attack on the cobs is reduced by cannibalism when larvae meet within the tight husks. Intercropping cowpea with sorghum can attract polyphagous pests onto the sorghum, which is a less valuable crop.

• The host-plant-finding behaviour of insects may be disrupted by the close juxtaposition of two plant species. Several crop pests, such as cabbage white butterflies (*Pieris brassicae* and *P. rapae*) and cabbage aphid (*Brevicoryne brassicae*) are very much influenced by the crop background in their colonization behaviour, and intercropping removes the contrast between seedlings and bare soil in the same way as dense planting does. Weeds, of course, have the same effect as a low intercrop; it has been shown that very few immigrant aphids were trapped over weedy plots of Brussels sprouts. Additionally, the mixture of odours from an intercrop, particularly any strong smell from a non-host plant masking the odour of the host plant of the pest, can disrupt the host-finding behaviour of pests. This has been shown, for example, by work in the UK at Cambridge, in relation to carrot fly (*Psila rosae*) on carrots interplanted with onions. There are many other, mainly anecdotal, records of aromatic plants repelling insect pests, particularly those of vegetable crops, and this is another area which is very amenable to a little experimentation.

- Intercropping may also increase the impact of natural enemies. This may be because one of the intercropped plants provides a honey or nectar source which attracts natural enemies for adult feeding, or because the shelter and humid conditions near the ground provided by the intercrop encourage ground-living predators. Work in the UK at Rothamsted Experimental Station, Harpenden, has shown that many predatory beetles are more abundant in weedy rather than clean plots of winter wheat. There is evidence that some ladybirds (Coccinellidae) prefer ground cover to rows of plants in bare ground. Some hover fly (Syrphidae) adults, whose larvae are important predators of aphids, also lay more eggs where the ground is covered than where it is bare; unfortunately other hover flies have the reverse behaviour. Another experiment at Rothamsted Experimental Station has endeavoured to establish aphid parasitoid populations by undersowing cereal crops with rye grass (*Lolium perenne*) and liberating an aphid (*Myzus festucae*), which lives on the grass but does not attack wheat, together with its parasitoid *Aphidius rhopalosiphi*. The parasitoid, however, also attacks grain aphid (*Sitobion avenae*), and there is some evidence that this procedure improved biological control of the aphid. Work on cabbage root fly (*Delia radicum*) of cabbages either grown traditionally or undersown with clover has shown that the clover undersowing greatly promotes the number of ground beetle predators of cabbage root fly eggs.

Separation of pest/vector and host in time and space

Attempting to separate the pest from its host plant in time or space is one of the oldest and most widespread farm practices often directly motivated by pest control, and it is still one of the most effective controls of some eelworm problems. Crop rotation normally reduces and delays attack rather than giving complete control because, although control may be significant within a given field, it is a less effective restraint over an area as a whole. Most pests have strong migratory powers or, if not, can frequently survive rotation on wild host plants. Moreover, crop rotation usually means that a particular crop is nevertheless grown somewhere close by in the area. Thus the common rotation for a field of grasses or cereals, followed by legumes and then root crops does not result in the absence of any of these crops on the farm as a whole. Yet the rotation is effective in reducing the many soil pests (e.g. wireworms [Coleoptera: *Agriotes* species], chafers [Coleoptera: Scarabaeidae], leatherjackets [Diptera: Tipulidae]) which multiply most successfully under grass. The various crop midges (e.g. pea midge *Contarinia pisi* and bladder pod midge *Dasyneura brassicae*) are weak fliers and also are disadvantaged by crop rotation, as is the carrot fly (*Psila rosae*) which has a flight range of less than 1 km. However, just to emphasize the point that cultural controls can often be a two-edged sword, it is worth giving the example of the wheat bulb fly (*Delia coarctata*), which strangely does not lay eggs in wheat crops, but in any fallow ground. The pest is therefore not a problem when cereals follow cereals, but only when cereals follow, through a rotation, bare fallow or a crop such as a root crop which leaves a fair amount of the soil surface exposed in late summer.

Crop rotation relies on the fact that there are usually only a few general feeders among the pests found across the rotation. For example, of 50 serious insect pests of the maize, wheat and red clover rotation, only three are important pests of all three crops. An equivalent to crop rotation in the field of veterinary entomology is 'pasture spelling', also known as 'pasture rotation' or 'host exclusion'. In this method for controlling ticks of cattle, potential host animals are excluded from pastures and the resultant lack of hosts means that ticks present amongst the vegetation die through lack of food.

It is necessary to understand the biology of ticks to appreciate the limitations of the strategy. Most ticks infesting cattle are so-called 'hard' ticks and belong to the family Ixodidae. When eggs laid on the ground hatch, they give rise to minute six-legged editions of the adults called larvae; these then moult to produce eight-legged larger nymphs, and these finally moult to give rise to adults. Now all three developmental stages feed on hosts such as livestock and wild animals. With a so-called 'one-host tick', the newly hatched larva climbs up vegetation and clings to a passing animal, such as a cow. It then feeds, moults to become a nymph which feeds on the same animal, and then the resultant adult also feeds on the same animal. So all three feeding stages remain on a single animal. In contrast there are 'two-' and 'three-host ticks'.

In a three-host tick, after a larva has blood-fed on its host, it drops to the ground eventually moulting to produce a nymphal stage which remains on the ground until it has the opportunity to climb onto another animal (same or different species) upon which to feed. The blood-engorged nymph drops to the ground, moults some days or weeks later to become an adult which then seeks out another host to feed upon.

Pasture spelling works best when the free-living stages, such as larvae and nymphs, are only short-lived when off the host, and when wild animal hosts are not within the pasture. In practice this means one-host species, such as those belonging to the genus *Boophilus*, are easiest to control, because with two- and three-host ticks nymphs and adults can survive a long time (6–12 months) on the ground off the host. The procedure is to remove livestock from selected pastures for two to three months and keep them isolated in small enclosures or paddocks, after which they can be returned to the pasture when the tick population should now be very small. In Australia alternating pastures every two to three months has worked well against *Boophilus microplus*. In Oklahoma, USA, when pasture spelling was operated on a rotational basis over 12 years, populations of *Amblyomma americanum* were reduced by 76%. When stock were excluded entirely from selected pasture, the reduction was 98%. In the UK the sheep tick, *Ixodes ricinus*, which transmits to sheep an arboviral infection called 'louping ill', has been effectively controlled by pasture spelling. In these two examples, in the USA and the UK, the ticks had a three-host life cycle which necessitates excluding livestock from pastures for many months to ensure tick populations have decreased substantially. Pasture spelling is sometimes combined with a minimal application of insecticides to the livestock.

Attempts to avoid plant pests by isolating crops from regularly infested sites are frequently designed to prevent insect-borne plant diseases from reaching the isolated crop. Because wild plants form reservoirs of both the insect vectors and the diseases they carry, the method has rarely proved successful on a regional scale.

Changes in agricultural systems

As emphasized at the start of this chapter, changes in agricultural practices may be purposeful to reduce insect problems or they may have been introduced for other reasons, in which case problems may as easily be found to have increased as decreased.

Purposeful variation of sowing date can control pests, most of which show some seasonal predictability. Thus it may be possible to avoid the egg-laying period of the pest or advance the age of the plants to a stage where they are resistant by the time the pest appears. For example, the hessian fly (*Mayetiola destructor*), has a predictable flight peak of limited duration; thus a few days' delay in sowing wheat can make all the difference between a good and a bad crop.

Another cultural pest control measure is early harvesting. This may remove pests (especially cereal pests in the straw and grains) from the field before they can emerge and perpetuate the population in the area. Damage to wheat caused by the wheat stem sawfly (*Cephus cinctus*) can be minimized by harvesting early (before the weakened stems lodge, i.e. fall over, in wind and rain) and much of the cultural management of cotton is designed to advance harvesting to before bollworms (*Helicoverpa* species) emerge from the plants as adults. Irrigation is terminated early and leaf desiccants may also be sprayed.

However, even changes designed to control insects may produce results that are not always beneficial. Planting crops that require no standing water instead of a second rice crop is sometimes promoted to reduce mosquito breeding. But in California, this practice has actually increased populations of the mosquito *Culex tarsalis* which is an important vector of Western equine encephalitis to humans. When rice fields were allowed to dry up after harvesting a rice crop, this killed the aquatic predators. Consequently when the fields were flooded again for rice cultivation, there were no predators to help reduce mosquito populations.

The potential of environmental/cultural changes to affect insect populations is perhaps most often revealed by changes introduced for reasons other than insect control.

A typical scenario in South-East Asia is farmers' houses sited near rice fields and pig pens, and people sitting outside their houses in the evenings. This is an ideal situation for the transmission of Japanese encephalitis. The rice fields support large populations of the vector, *Culex tritaeniorhynchus*, while humans and pigs are attractive hosts for these mosquitoes. In addition pigs are amplifying hosts, i.e. they are animals in which the virus of JE multiplies enormously. Therefore, when a blood-meal is taken by a mosquito a large amount of virus is ingested which leads to increasing the chances of transmission when the mosquito feeds again, possibly on humans. But there have been changes in agricultural practices and economic conditions. For example, in Japan when in the 1950s morbidity from JE was highest the pig to man ratio was less than it is now. The increase in the pig to man ratio has been accomplished by a *decrease* in the number of pig farms but accompanied by an exponential increase in pig numbers on farms as they have become much larger. Other factors are that farmers are living further away from piggeries and rice fields, and people are no longer sitting outside their homes drinking saki in the evenings but sitting in screened houses and porches, and they also sleep under mosquito nets. This has led to a decline in JE in humans, although there have been resurgences of pig epizootics. This serves as a good example of how socio-economic improvements have led to a decrease in disease transmission, albeit not intentionally.

Two changes since the 1960s, the move from spring to winter wheat and changes in sowing sugar beet, provide striking examples from agriculture as to how changes in agronomic practices have caused new pest problems.

The increased growing of winter wheat, sown even earlier to exploit non-tillage systems, has created a whole range of new problems related to the changed timing of the crop rather than stemming just from the reduction in soil disturbance (see earlier). A stem-boring fly (*Opomyza florum,* which only merited a 'common name', the grass and cereal fly, once it had become a problem) now finds cereals at the right stage of growth when it lays its eggs in late summer. Before winter wheat, it had to manage with the poorer food resource of wild grasses. Also, winter cereals are infested by the aphid *Rhopalosiphum padi* in the autumn and can be infected with barley yellow dwarf virus from wild grasses by this aphid. The virus then multiplies in the plants for the rest of the crop season and can cause severe symptoms the following year. Spring wheat avoids this autumn aphid infestation and, although the virus may be brought in by other aphids much closer to harvest, the virus is then not much of a problem.

Sugar beet seed used to be polygerm (i.e. each seed produces several seedlings), and was sown thickly and later thinned to the final desired plant population. Recent changes in crop management have included precision drilling to plant stand for sugar beet, made possible by the development of monogerm seed. Attack on seedlings of these crops always occurred, but seedling losses were compensated by the dense stand of seedlings which always necessitated laborious and costly thinning. Now seedling pests such as pygmy beetle (*Atomaria linearis*) and mangold fly (*Pegomya hyoscyami*), which used to be regarded as unimportant pests of sugar beet, have become problems. A similar change to precision drilling of cereals has equally put a greater emphasis on seedling pests.

Conclusions

It may often be possible to conceive of a cultural or environmental change which could cause considerable reductions in a problem insect. The premium that mankind puts on life itself means that people might accept considerable inconvenience with an environmental control method for a serious disease vector if it really was effective. Relatively few crop insects are in the same league. It is therefore much harder to come up with a plant cultural change which would also be acceptable to the farmer, particularly in developed agriculture. However, agronomists and cropping systems scientists in many institutions are continually setting up experiments of new cultural systems. They do this in the belief that these systems might gain farmer acceptance for economic and agronomic reasons. It may well be just as sensible for the worker interested in pest control to take data from the experiments of these colleagues than to go it alone and try to develop new systems purely for pest-control purposes. At the very least, by now history should have taught us that changes in the way mankind does things can cause new pest explosions (see also Chapters 1 and 13).

7

Biological control

Introduction

One of the oldest methods of pest control is the use of other animals as carnivores to reduce pest numbers. More recently other biological organisms, such as microbes causing insect diseases, plants which are resistant to pest attack and animals that tolerate, at least to a certain extent, infection with diseases, have been used for pest and disease control. These are discussed in succeeding chapters, while 'biological control' in this chapter is limited to the use of animals that are natural enemies of insects. There are no elementary recent texts on biological control; the slim volume by Samways (1981) can still be recommended. An excellent recent book, but an advanced text (more than 500 pages), is that by Van Driesche and Bellows (1996).

History of biological control

The use of biological control is probably about as old as the history of agriculture. Chinese cave paintings clearly show ducks being used to consume pests at the base of plants in rice paddies, a technique still in use in China today. However, the first well documented case of biological control occurred in 1762, when a Mynah bird (*Gracula religiosa*) was brought from India to Mauritius to consume locusts. In the 1770s, the practice developed in Myanmar (Burma) of creating bamboo runways between citrus trees to enable ants to move between the trees more freely for the control of caterpillars.

However, the first real landmark in modern biological control dates from 1887, when a ladybird (*Rodolia cardinalis*) was used to control a scale insect (*Icerya purchasi*) on citrus in California. So much is this regarded as a starting point, that 1987 was chosen for a major international conference to celebrate '100 years of biological control'. This Californian example of biological control will be discussed in more detail later, but it led to a rush of activity exploring the possibilities of biological control in the early twentieth century. Some schemes were rather far-fetched, such as building in Lousiana, USA, bat towers in the belief they would provide roosts for insectivorous bats, and thus reduce populations of local nuisance mosquitoes. Already by 1935, 26 successful examples of biological control could be listed, but soon afterwards the introduction of DDT produced such easy, cheap and successful control that there was considerable disillusionment with biological control which, for most people, was no longer seen to have much value. For instance, many, but not all, unrealistically believed that DDT could lead to global eradication of malaria. But the euphoria of the DDT era did not last long, and by 1958 the 26 examples of 1935 had only grown to 100. The problems created by insecticides, such as environmental contamination, evolution of insecticide resistance, and even sometimes increased pest populations, soon led to renewed interest in biological control, and by 1964, 225 examples of successful or partially successful biological control programmes could be quoted. Today there are probably about 500 examples, though it must be pointed out that an 'example' is a local success in biological control, and may well duplicate a system previously employed elsewhere. Nevertheless, probably about 150 species of pest are involved in the successes of biological control reported to date. About one in eight biological control programmes prove successful.

Renewed interest in biological control against medical vectors came a little later, in the 1960s and through the 1980s, than for crop pests. In 1967 the World Health Organization in Geneva established a Scientific Working Group on Biological Control of Vectors to promote research and application of biological control methods. Nevertheless despite this, nearly all the examples of successful biological control programmes have been against agricultural and horticultural pests. There have been very few successes in controlling medical or veterinary pests or vectors (see p. 170). There are several problems that apply particularly to medically important pests. Whereas much successful biological control of plant pests has been against exotic pests, most medical and veterinary pests are indigenous. Also unlike pests of plants, their environment is not the homogeneous one of field or greenhouse monoculture, but a heterogeneous one often leading to a very patchy distribution of the insects. Moreover, although annual crops are not permanent and the feeding niches within them are even more ephemeral, medically important vectors often breed in exceptionally transient and certainly less predictable habitats, such as ground pools, water-filled tree holes, tin cans and small streams. All of these are liable to desiccation, when both pests and biological control agents are killed. Whereas the pest species (e.g. mosquito) can rapidly recolonize, biological control agents do not; consequently pest outbreaks often result.

Biological control has many advantages as a pest control method, particularly when compared with insecticides. However, there are also serious disadvantages limiting its popularity with growers and control operators and the situations where it can be used.

Advantages of biological control

(a) *The technique is selective with no side effects*. Biological control agents tend to be fairly prey-specific, and obviously do not carry the kind of environmental dangers of broad-spectrum kill associated with insecticides. This does not mean that side effects can be totally excluded, although they have been very rare in the history of biological control. One of the most notorious examples of such side effects was the introduction of the cane toad, *Bufo marinus*, to Australia to control the cane beetle, *Dermolepida albohirtum*. The toad was brought in by farmers in 1915, very much against scientific advice. The toad spread rapidly; unfortunately it also fed on many small animals (especially frogs) and so snakes were driven to extinction or near extinction; also the toad is very poisonous, and killed many carnivores, the bodies of many of which were found with toads still in their mouths. On another occasion a serious disease of sugar cane was introduced into Trinidad on the ovipositor of a parasitic insect being brought in for biological control purposes. Also there have been at least two cases where, after controlling the intended prey, biological control agents have switched to other related herbivores which were important in controlling weeds. Some kinds of side effects on other insects are almost inevitable. The success of the parasitoid used against cassava mealybug (*Phenacoccus manihoti*) in Africa (see below) has resulted in a decline of the local ladybird (Coccinellidae) predators, because the parasitoid caused such a dramatic reduction in the total prey availability in the area.

(b) *Biological control is non-polluting*. Clearly biological control does not introduce pollutants into the ecosystem/environment.

(c) *Biological control can be cheap.* It rarely costs more than about US$ 3 million (as compared with US$ 100 million or more to develop an insecticide) to complete a biological control exercise, and moreover, in classical biological control, which involves establishing exotic natural enemies, this cost often has to be met only once. Furthermore, such classical biological control is usually free of charge as far as the user is concerned. Thus it is a particularly attractive option for some forestry and pasture problems, and for many tropical crops grown which have low inputs and are unable to carry the cost of an insecticide. In such situations, biological control does not have to stand comparison with the levels of control given by insecticides, because it may be the only economic solution.

By contrast in terms of cost, the inundative release of natural enemies in glasshouses or on crops involves repeated purchases of insects from specialist suppliers, who have to staff and maintain expensive insectaries and have to rear both the natural enemies and their prey (the pests). Thus it is not unusual for biological control in glasshouses to cost eight or nine times the equivalent control obtained with insecticides. By 2002, more than 100 species of natural enemies were commercially available for use in glasshouses.

(d) *Biological control agents have the potential to be self-propagating and self-perpetuating.* Ideally, once introduced, biological control agents will persist in time, and may spread over large areas from the points of release and reach targets that chemicals cannot (such as larvae concealed in fruit, stems or underground, or tsetse fly puparia hidden in the soil). This clearly has occurred where they have co-existed with their prey for centuries, but the ability to self-perpetuate may not be realized in a new environment for a variety of reasons, including climatic constraints and lack of alternative prey (see below).

(e) *The development of resistance of pests to biological control is unlikely.* Insects are often capable of defence against attack by carnivores and may, for example, exhibit escape behaviour, release repellent chemicals or encapsulate and suffocate foreign bodies such as parasitoid eggs or nematode parasites. Mechanisms for resistance which could be increasingly selected for by strong pressure from biological control agents in an analogous way to resistance to insecticides are thus already present in the population, and much has been written about the potential danger this presents for long-term biological control. However, an existing natural enemy of a pest has clearly already adapted to such mechanisms and is, moreover, capable of further adaptation; indeed, we know of no cases where previously successful biological control has failed because of selection for resistance.

Disadvantages of biological control

(a) *Biological control limits the subsequent use of pesticides.* Where biological control agents are being used against one pest, it is clearly difficult to continue using insecticides against other pests on the same crop or other disease vectors in the same area. This may make the use of biological control impossible unless biological control systems can simultaneously be set up against other pest insects (see p. 274). Many people assume that the use of pesticides against the prey of the biological control agent is similarly limited. Indiscriminate pesticide use is certainly so limited, but (see Chapter 13) some ingenious ideas for using broad-spectrum pesticides have enabled the latter to be the key to making biological control of a pest more effective by making kill more selective in favour of the natural enemies.

(b) *Biological control acts slowly.* It obviously takes some time for biological control agents to spread from their points of release, to build up numbers and to make their impact on the pest population. Some experienced biological controllers believe in a 'rule of thumb' of eight generations, which of course is eight years if the pest is univoltine. During this period, when the pest may still be present at intolerable levels, any use of pesticide against it or other pests on the crop can endanger the biological control system. The need for a difficult transition period from insecticides to biological control may make the introduction of biological control unattractive for a grower. Moreover, in disease epidemic situations, such as dengue or malaria outbreaks, it is clearly impossible to allow infection to continue while biocontrol agents take effect. The only practical solution is control with insecticides, such as aerial spraying to immediately kill infected vectors.

(c) *Biological control rarely exterminates the pest.* A biological control system, if intended to be self-perpetuating, involves the continued presence of the prey, even if only at low levels. Growers cannot therefore expect to have a totally clean crop as they can with insecticides, and there may be several types of pests (e.g. blemishers of quality, disease vectors) which, even at low levels, will still cause economic damage. Similarly, although in the long term biological control agents may reduce disease transmission to humans or livestock to so-called 'acceptable levels', people then often still aspire to disease eradication.

(d) *Species-specificity.* Scientists see the specificity of biocontrol agents as a virtue in developing more target-specific controls. Indeed, non-specific agents are usually rejected for release in classical biocontrol programmes. Considerable work and money may go into developing a control programme that gives control of just a single pest species. However desirable this is in pest management and environmental terms, the species specificity of most natural enemies makes the economics of producing insects as biocontrol agents unattractive to commercial companies. For example, in 1980 there was just one commercial company in the USA interested in producing parasitic nematodes (*Romanomermis culicivorax*) to kill mosquito larvae, but it then decided that the market was too small for a profitable return and abandoned the project. Many biocontrol companies have come and gone, and take-overs and amalgamations are frequent.

(e) *Biological control can be unpredictable.* Growers have relatively little control over a biological control system once it is in place, and this often may worry them. Even working programmes can suddenly fail. The ladybird *Chilocorus cacti* was released against mulberry scale (*Pseudaulacaspis pentagona*) in Puerto Rico in 1938. The control was successful, but over a long period the ladybird virtually exterminated the scale and then died out. A sudden mass outbreak of scale recurred in 1953. Similarly, in the biological control of whitefly (Aleyrodoidea) in glasshouses, a sudden change in weather or a period of extreme hot or cold can cause a breakdown of the system. Usually greater knowledge and biological expertise are needed to initiate and sustain biological control programmes than insecticidal ones. Insecticides cause density-independent mortalities which should be predictable, whereas most biological control agents cause density-dependent mortalities and this can make the outcome unpredictable. Particularly, biocontrol other than that achieved by purchasing large numbers of biological control agents from commercial insectaries for inundative releases is not under the grower's control and operates over a larger area than one holding; the grower has to hope and trust the scientists, but not infrequently worries.

(f) *Mass rearing and transportation.* There may be difficulties in mass rearing, maintenance, survival, storage and transportation of biological control organisms. It is not unknown for variability in 'batches' for inundative releases to occur; a batch of eggs parasitized by the wasp *Trichogramma* may show less that 20% emergence of adults. Although a reputable company can be expected to replace the batch free of charge, the time lost may prove a commercial disaster for the grower.

The range of animal biological control agents

Predators

These are perhaps the best known agents. They capture, kill and consume numerous prey individuals during their development. They include birds, bats and other larger animals for terrestrial pests but aquatic predators, especially fish, for insects with aquatic stages in their life cycle e.g. mosquitoes. However, with crop pests the great majority of predators are arthropods such as spiders, mites and insects. Many orders of insects have representatives which are predatory. Those used in biological control are predominantly heteropteran bugs, beetles, flies, lacewings (Chrysopidae) and predatory mites. Other insects (e.g. earwigs and predatory wasps) have only occasionally been used.

Sometimes, as with ladybirds, both larvae and adults are predators; otherwise, as with many predatory flies like hover flies (Syrphidae), only the larvae are carnivorous.

Monophagous predators that prey on just one species or group of closely related species of host are rare. They have sacrificed the ability to prey on a range of hosts to specialize in seeking out a single species, or more commonly species having similar behaviours. A polyphagous species feeds on a range of diverse species, although often they are not as efficient at catching a specific species as monophagous predators and, if released for biological control out-of-doors, could pose a threat to non-target organisms. The usual compromise is an oligophagous predator, which is generally better than either monophagous or polyphagous ones for biological control. This is because although it may favour one type of host, hopefully the target pest, when this becomes scarce it can switch to feeding on alternate prey and so avoid starvation. When the pest starts to become more common again, however, it switches back again to feeding on the target organism. This has a stabilizing effect on predator–prey relationships. Oligophagous predators generally also respond better to changes in pest population densities than do polyphagous ones.

Parasites and oligophagous predators commonly have poor searching abilities, quite the opposite of parasitoids (see below). Predators often selectively prey on diseased individuals because their escape reactions are not very good, but this is detrimental to biological control because such prey are probably already doomed to die. Most arthropods have highly aggregated (contagious) distributions within their habitat. It probably benefits predators, and parasites, not to randomly search for prey but to concentrate their attacks where the prey are aggregated. Clearly this leads also to aggregation of predators and, as the prey become scarcer through predation, mutual interference amongst the predators will ensue. The best strategy then is for the predator to seek out prey in less favoured areas having smaller numbers of prey, but also fewer predators. It is obvious that using predators for biological control is ecologically complex and it can be exceedingly difficult to predict the outcome.

Parasitoids

Insect parasitoids are found mainly in the Terebrantia group of the Hymenoptera and the family Tachinidae in the Diptera. There are a few other parasitoids in other orders (e.g. Coleoptera, Strepsiptera). Parasitoids comprise about 10% of all known insect species. They are usually host-specific and have very efficient searching abilities. In contrast with predators, each individual parasitoid completes its development using one prey (i.e. host) individual, which is killed in the process. With many parasitoids the prey hosts the development of a single parasitoid, even if several parasitoid eggs are laid on or into the host (superparasitism). With other parasitoids, many individuals may share one host.

Another distinction is between endo- and ectoparasitoids. Endoparasitoids lay their eggs in, and the larvae develop within, the host (Fig. 7.1), whereas ectoparasitoids lay their eggs on the outside of the host and the larvae remain on the outside, usually with their mouthparts thrust inside the host.

The Terebrantia are the parasitoids most widely used in classical biological control. The group includes both endo- and ectoparasitoids. The Tachinidae have been far less exploited for biological control; ectoparasitism is the norm. Unfortunately there are no known parasitoids of mosquitoes, and only very few attack simuliid black flies. They more commonly attack medically important arthropods that do not have an aquatic life stage, such as triatomine bugs, muscid flies and tsetse flies. Although species of the hymenopteran genus *Nesolynx* have been used against tsetse flies, results were disappointing.

Nematodes

Nematodes are also used in biological control, and are more like some parasitoids than predators in that a single host will harbour many nematodes and

Fig. 7.1.   The parasitoid wasp *Aphelinus flavus* laying an egg into the aphid *Aphis gossyppii,* a pest of cucumbers. The parasitoid larva will develop inside the aphid, killing it before pupating within it and finally emerging as an adult through a hole cut in the skin of the dead aphid (courtesy of Horticulture Research International).

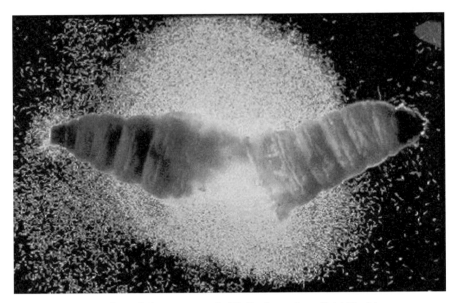

Fig. 7.2. Caterpillar of the wax moth (*Galleria melonella*) killed by a nematode (*Steinernema carpocapsae*), showing the vast numbers of migratory J3 larvae produced to infect new hosts (courtesy of N. G. M. Hague).

will die as a result of the attack (Fig. 7.2). With the majority of nematodes, death is not the direct consequence of nematode feeding, however, but of a toxin released by bacteria that are carried into the insect host by the nematode, and which cause breakdown of the gut wall leading to death by septicaemia (cf. *Bacillus thuringiensis,* Chapter 8). Nematodes are drought resistant and sufficiently small to be amenable to storage and spray application in much the same way as pathogens. Comparatively little use has been made of nematodes in this way, although they have been applied against the Colorado beetle (*Leptinotarsa decemlineata*) in Canada; at the moment there is a great deal of research on one particular genus, *Neoaplectana. Neoaplectana carpocapsae* has been shown experimentally to be effective against codling moth (*Cydia pomonella*), an apple pest. Nematodes (*Steinernema feltiae*) marketed under the appropriately indicative name 'Nemasys' have proved particularly useful in controlling a number of soil pests (e.g. vine weevil, *Orthorhinus klugi*) against which the previously effective insecticides are no longer permitted. They also may be the only possible biocontrol agents for several important pests (e.g. the mushroom fly *Sciara* and thrips).

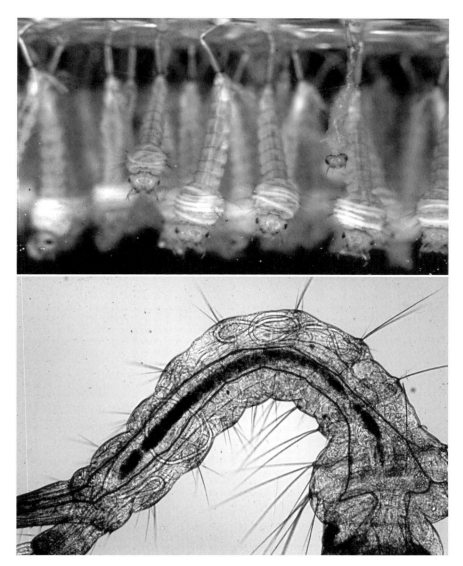

Fig. 7.3.   Mosquito larvae showing mermithid nematodes, *Romanomermis culicivorax*. Top, coiled inside their thoraces; bottom, 3–4 larvae in the abdomen (courtesy of E. G. Platzer).

Some years ago there was considerable interest in the aquatic mermithid parasite *Romanomermis culicivorax* (Fig. 7.3) which infects mosquito larvae, and several trials were undertaken. Unlike most other insect-pathogenic nematodes, mermithids are quite large and kill by their feeding. A commercial product was made called 'Mosquito Doom'. This nematode is highly pathogenic and highly infectious, with infection rates of 84–97%. Such high mortality rates lead to the parasite dying out because all its hosts are killed, thus necessitating repeated introductions. This and the fact that the worms became very aggregated in ponds and showed little dispersal which meant there had to be many application points, as well as its high specificity to just mosquito larvae, resulted in cessation of any commercial production.

This nematode provides a good example of density-dependent population regulation. When mosquito larvae are parasitized with three or more nematodes the worms mature only, or predominantly, into males, which puts a brake on sexual reproduction and thus reduces the number of mosquito larvae infected and killed, necessary if the nematodes are to survive.

The techniques of biological control

Inoculation

This is often called 'classical' biological control, as it was the approach taken in the early examples of biological control against imported pests. The natural enemy is liberated in relatively small numbers in the hope that it will establish itself. This approach has been used particularly, though not exclusively, for the control of imported pests, but it could equally follow a vacuum in natural enemy activity created after spraying. It has proved especially useful on perennial plants, against sedentary pests and in 'ecological islands'. For inoculation to be an applicable technique, it is important that the pest should not cause damage at low densities and be known to have arisen as an insecticide side effect, or to have been introduced without its natural enemies. Inoculation is also particularly applicable where the problem is widespread and the crop needs little insecticide against other pests.

In the early days of classical biological control, releases of agents and their success or failure were largely hit-or-miss affairs based on the collection abroad of many different potential agents and their release with little preliminary scientific evaluation, such as the widespread introductions of larvivorous fish, *Gambusia affinis*, in the hope that they would achieve mosquito control. Today the programme is much more deliberate, and follows a clear sequence.

Firstly, a survey of possible natural enemies is often carried out in that area of the world assumed to be the centre of evolution of the pest species. It is often hard to identify which natural enemy abroad is likely to be the most effective back at home. This is because an agent capable of matching its prey in reproduction and so able to suppress outbreaks may be quite rare in a situation where outbreaks are absent or infrequent. Here the most abundant natural enemy may not be of much use when transferred from an endemic low prey population to an outbreak situation in a different country. However, there is sometimes merit in seeking a new type of attack; for example, if natural enemies are mainly attacking larvae at home, then a pupal parasitoid found abroad may well be effective. Also, there might be particular hope for a biological control agent which is closely related to a type which succeeded elsewhere in the world on a similar pest problem.

It is therefore usually necessary to send a range of natural enemy species back to the home country, where they are kept in strict quarantine conditions to eliminate diseases and other sources of mortality as far as possible. This will give them a potentially 'unnaturally' high survival rate after release which may often improve the impact they make on the pest. They will be reared in cages, often on unusual food, and there is usually a decline in the imported population before numbers begin to build up to the point where release is contemplated. Since a relatively small sample will have been imported, and then forced through the genetic bottleneck of culturing, the gene pool of the released individuals will be very small compared with that of the parent population abroad. This would suggest that the released population is likely to be rather unadaptable, and this may account for several failures of establishment that have occurred following release. However, should the released population establish satisfactorily, the reduced gene pool may actually be an advantage in limiting the ability of the pest and natural enemy to co-evolve towards maintaining a higher pest equilibrium level.

During the quarantine phase, a number of studies will be carried out on the selected agents to assess their potential suitability in the field. In general it is desirable that the species should have a high searching capacity (e.g. parasitoids), so that they will not emigrate readily when host numbers decline. A host-specific species is likely to be more effective at low pest densities, but will be more prone to seasonal shortages of prey or to ecological interferences such as harvest of the crop. A high reproductive rate is desirable, and is particularly useful if, like many small parasitoid wasps, the species has asexual reproduction. It is also important that the species is adapted to the range of climatic variation it is likely to encounter in the field.

As well as studying climatic adaptability, it is also important to study the relationship of the development and voracity of the natural enemy to temperature. This will determine whether the biological control agent can cause mortality sufficiently early in the pest's annual cycle to be effective and whether it can then subsequently avoid being 'outstripped' by the pest. It may even be worth looking for natural enemies in a cooler area than the home country in the hope of finding an ecotype with a low development threshold to advance the time of its appearance in the pest population during the season.

Tests also need to be carried out to determine how safe it would be to release the natural enemy in the field, in relation to possible alternative prey, which may already be beneficial as biocontrol agents of another insect or particularly a biocontrol agent for a weed. For instance the top minnow, *Gambusia affinis*, is a voracious predator and originally was a native of south-eastern USA, eastern Mexico and the Caribbean. But it has been introduced into more than 60 countries to control mosquitoes, consequently making it the most widely distributed fish in the world! However, because of its pugnacious character it has reduced, or eliminated, several indigenous fish species and in 1982 the WHO concluded that this fish should not be recommended for introductions. In addition *Gambusia* preys on invertebrates including those that graze on phytoplankton, and this has sometimes resulted in algal growth, which may cause these fish, and others, to die, thus allowing mosquito populations to increase. Increasingly the value placed on biodiversity means that efforts are made not to release agents which might affect even non-targets of no apparent economic value.

To establish a natural enemy in the field involves trial releases in a number of contrasting habitats; usually 1000 to 5000 individuals are released annually over a 5-year period. The success of the release then has to be monitored over a period of several years to observe whether the pest is declining and whether natural predators are attacking the biological control agent heavily.

If, after a number of years, only partial biological control is observed, it should not be abandoned. Instead, studies should be undertaken to see whether the biological control agent can be supported by a secondary measure (such as cultural control or the introduction of a plant variety less susceptible to the pest). Increasingly there is interest in exploiting even partial biological control, integrated with other control measures (see Pest management, Chapter 13).

Seasonal inoculative release

A special use of inoculation has been attempted in glasshouses, particularly to control red spider mites (*Tetranychus urticae* and *T. cinnabarinus*) by the predatory mite *Phytoseiulus persimilis* (Fig. 7.4). This is based on the concept that the natural distribution of the pest in localized aggregations is disadvantageous to biological control (see earlier). First the pest is inoculated evenly across the crop so that it is already present when the predator is

Fig. 7.4. Predatory mite (*Phytoseiulus persimilis*), a biological control agent for glasshouse red spider mite (*Tetranychus urticae*) (Holt Studies Picture Library).

inoculated, similarly evenly but after a suitable time interval. This procedure ensures that predators are present wherever pest spider mites arrive in the glasshouse. Even if biological control is not perfect, it may considerably delay the time it takes the pest to reach damaging levels, and the whole system can be wiped out with an insecticide if, as often, biological control finally fails. There is probably no reason why the technique should not also be attempted in annual crops outdoors.

Inundation

This is the use of biological control as a biological pesticide! Large numbers of the natural enemy are reared in the laboratory and liberated onto crops; indeed, there are a number of companies culturing and selling biocontrol agents for this purpose in glasshouses. The aim is to create an outrageously high ratio of biological control agents to pests so that the pest is exterminated, the biological control agent itself dies out, and pesticides can then safely be used against other pest species. It is a technique which has particular appeal where a pest population has become resistant to the available insecticides.

An early use of inundation, beginning in 1960, was against black scale (*Saissetia oleae*) in Zanzibar. The programme involved one to three inundative releases per year of the encyrtid parasitoid wasp *Metaphycus helvolus*.

In California, predators of aphids have been persuaded to inundate themselves into lucerne fields. The natural enemies are attracted into the fields by an odour emanating from aphid honeydew, and therefore normally do not invade the fields until large aphid populations have developed. We now know that the principal attractant is indole acetaldehyde, a breakdown product of the amino acid tryptophan in plant sap and therefore in the honeydew. In the terminology of Chapter 10, indole acetaldehyde is acting as a 'kairomone', but the response of at least some natural enemies first has to be triggered by the odour ('synomone') of host plants of the aphid. However the odour from indole acetaldehyde can be mimicked by spraying the crop with a waste product from brewing, a yeast hydrolysate called 'Wheastrel'. This early example of manipulating natural enemies by their responses to chemicals which influence their behaviour has recently developed new momentum as understanding of behaviour-modifying chemicals has improved (see Chapters 10 and 12).

In the USA, and several other parts of the world, inundative releases of the large and colourful mosquitoes of the genus *Toxorhynchites* have been made in the hope that their large predatory larvae will control the container-breeding *Aedes aegypti,* vector of dengue virus, and in Africa and South America also yellow fever. Despite repeated mass releases of laboratory-reared adults and some temporary control the method is not sustainable and logistically cannot be used over wide areas.

Conservation

Today there is ever-increasing emphasis on maximizing the activity of indigenous natural enemies by either avoiding their large-scale destruction when insecticides are used or by improving the environmental conditions to enhance their survival and activity. There are thus two quite separate approaches to Conservation Biological Control. It is unfortunate that the use of exactly this terminology (Conservation Biological Control) has recently surfaced as an umbrella for the second approach only. This approach, previously given the more appropriate and descriptive title of habitat modification is discussed here, while how careful and even ingenious insecticide use can conserve biological control is dealt with in Chapter 13. It will become clear that often only one plant species needs to be added and that this will often be more effective than trying to mimic naturally evolved biodiversity. However, it will also become clear that often the motive for farmers to include some floral diversity in their management plans – diversity which is very likely to benefit biological control – is quite different. It is probably fair to say that such diversity without a biological control motive has been taken up far more rapidly and widely than the introduction of diversity planned specifically for biological control.

Microclimate and crop background

There have been many reports that high humidities favour biological control. In coffee, the pruning system is often designed to maintain high humidity at the right time for parasitoids of the serious bug pest *Antestiopsis*. Several sources also report that enhanced natural enemy activity in weedy crop plots as opposed to clean ones has a humidity explanation. It may be difficult to envisage the practical feasibility of encouraging farmers to leave their plots weedy, but nevertheless some cereal farmers in the UK have taken up the concept of leaving an edge area of the crop unsprayed by insecticides and broad-leaved herbicides. This is easily accomplished by switching off the outside half of the spray boom when the tractor is spraying along the field edges (Fig. 7.5). Although this has been done mainly to provide insect food for game birds on farms, it seems likely that the technique will lead to an increase of the natural enemy restraint on insects on the central area of the crop.

The importance of appropriate microclimatic conditions for the overwintering of predatory ground beetles has been exploited with the concepts of 'beetle banks'. A simple ridge of soil sown with a tussocky grass such as cocksfoot (*Dactylis glomerata*) and fitting under the post and wire strand fencing around cereal fields will produce huge numbers of overwintered beetles to move into the fields in the spring. In large fields, such ridges can be created at 150-m intervals (Fig. 7.6). Sadly, although these techniques have been shown to be economically viable, uptake by farmers has been negligible. However, financial incentive schemes in the UK such as 'set-aside' and 'countryside stewardship' are encouraging farmers to benefit birds and other elements of biodiversity with areas in their farms not dissimilar to beetle banks and conservation strips (see later); these must prove beneficial for biological control also.

Fig. 7.5. Selectively spraying of cereal fields margins. Above, the tractor with the outside boom shut off while spraying along the edge of the crop; below, the floral diversity created thereby (courtesy of N. W. Sotherton).

Alternative and alternate prey

Natural enemies may die or emigrate if they reduce their host population to very low or zero levels. Although we have argued above that it is an advantage for a biological control agent to be very host specific, so that it does not attack non-target prey, we also pointed out that sometimes the presence of other insects as alternative prey is necessary to make biological control

Fig. 7.6.   Raised ridge sown to the grass *Dactylis glomerata* running across a cereal field (courtesy of S. D. Wratten).

self-perpetuating. For example, gypsy moth (*Lymantria dispar*) outbreaks tend to occur in forests without ground vegetation. Where there is ground vegetation, however, such flora support many caterpillars of other species, which are used by parasites of the gypsy moth at times when the gypsy moth is scarce. The provision of alternative prey is one way in which biodiversity schemes in farmland (see above) are likely to encourage natural enemies of insect pests.

There is of course the danger that alternative prey in non-crop areas may arrest the natural enemies and that they will then not move into the crop. For example, ladybirds often feed on aphids on shrubs and trees in the spring, and do not move onto farm crops in time to control aphids there before large populations have developed. This is one aspect that suggests a potential for using behaviour-modifying chemicals (see the example of Wheastrel given earlier, and Chapters 10 and 12). Another possibility that was tested, but not taken up in farm practice, was to undersow wheat with rye grass (*Lolium perenne*). This is colonized by an aphid species which does little damage to the grass, but is a host for a parasitoid which will move onto wheat aphids (*Sitobion avenae*) when these later appear on the cereal (see p. 141).

Alternate prey is prey that is required by the natural enemy in order to maintain its survival regardless of the abundance of the pest species. A good example here is the grape leafhopper (*Erythroneura elegantula*) in California. Its egg parasitoid, *Anagrus epos*, requires a generation of leafhopper eggs for its overwintering generation, but unfortunately the grape leafhopper itself over-winters as an adult. Biological control of grape leafhopper by *Anagrus* occurred where blackberries (*Rubus fructicosa*) were growing near the vineyards. Here the blackberry leafhopper (*Dikrella californica*), which overwinters as an egg, acts as a bridging host for the parasite throughout the winter season. It was later discovered that other leafhoppers, especially *Edwardsiana prunicola*, on trees growing like the blackberry near watercourses, were rather more important sources of the parasitoid.

In Great Britain, the diamondback moth (*Plutella xylostella*) is heavily par-asitized by *Diadegma* species. The parasitoid emerges from caterpillars on cabbage in the autumn, whereas the caterpillars themselves spin up as a cocoon for the winter and are not suitable as hosts for an overwintering genera-tion of the parasitoid. It has been known since the 1930s that *Diadegma* must bridge the winter in some other caterpillar, but it was not until the 1950s that the late O. W. Richards located it overwintering in a caterpil-lar (*Swammerdamia lutarea*) on hawthorn (*Crataegus monogyna*). *Swammer-damia* was, until then, just another 'economically neutral' insect, and farm-ers had been actively uprooting their hawthorn hedges to enlarge arable field sizes, oblivious of the fact that this plant was a vital link in the bio-logical control of a potentially very damaging pest of cabbages and other crucifers.

Flowers

Flowers are important sources of food for adult insects. The abundance and diversity of insects visiting flowers, especially the large white platforms of flow-ers like hedge parsley in the family Umbelliferae, are often exploited by insect collectors. Many adult insects have to feed on pollen and/or nectar before their eggs can mature. Such insects include both pest and beneficial species: it has been calculated that eight cow parsley plants can feed at least 2000 cabbage root fly (*Delia radicum*) adults between emergence and oviposition. Flow-ers have similar attraction to many natural enemies, particularly the para-sitoids. Russian workers have found it advantageous to place potted flowering Umbelliferae in cabbage crops at the ratio of one pot to 400 cabbages in order to promote biological control. In the absence of flowers, especially since the use of extremely effective herbicides have eliminated weed flowers in most crops, natural enemies emerging as adults may have to leave the crop to find food before their eggs will mature.

The provision of flowers as nectar sources near crops is today seen as a major aspect of habitat modification. Several plants, especially American buckwheat (*Phacelia tanacetifolia* in the family Hydrophyllaceae), seem to be especially attractive to natural enemies while less attractive than Umbelliferae to most pest species. Planting such plants at field edges or on the beetle bank ridges within cereal fields (see earlier) creates valuable 'conservation strips'.

Wine producers in Switzerland have been driven to try Conservation Biological Control, as insecticides approved for vineyards have almost all been withdrawn because of the reluctance of industry to pay for re-registering their older compounds against newer, much stricter criteria. Carrying out the research to satisfy these new criteria involves huge costs. In desperation, the vine growers have sown flowering 'mini-meadows' in the spaces between the rows of vines (Fig. 7.7). The level of pest control this has given, probably because of the provision of alternative hosts as well as flowers, has exceeded all expectation, certainly in the damper areas.

Some examples of successful biological control

Examples from crop situations

Cottony cushion scale in California
We tend to think of resistance to insecticides as a fairly recent phenomenon, but by 1887 cottony cushion scale (*Icerya purchasi*) had developed resistance to the insecticides then available, culminating in the failure even of pumping hydrogen cyanide gas under canvas covers put over the bushes. The full story of this biological control programme has many peculiar facets, such as an unfortunate love affair, irregularities with government money and a pair of diamond earrings. The account by Doutt (1958) is well worth reading. By 1888, an American scientist had located two potential biological control agents in Australia. One was a tachinid fly and the other the ladybird *Rodolia cardinalis*. The Americans had considerably more hope of the tachinid than the ladybird, and so the first shipment of insects from Australia comprised 12 000 flies and just over 100 ladybirds, though another 380 of the latter were sent later. However, it was the ladybird (Fig. 7.8) that gave complete success only 15 months after arrival of the first shipment, and it has since been used in many other places against the same pest.

Fig. 7.7. Flowering mini-meadows between vines in Switzerland (courtesy of D. Gut).

Coconut moth in Fiji

This example is cited for several reasons. It is another early example, but also illustrates the successful use of the kind of biocontrol agent which was not successful on citrus in California, a fly in the family Tachinidae; it is cited also because it involved a new association, i.e. the transfer of the biological control agent to a new host. In the 1920s, British entomologists imported the tachinid fly *Bessa remota*, which normally parasitizes another moth, to

Fig. 7.8. A famous success in biological control: the vedalia beetle (*Rodolia cardinalis*) feeding on cottony cushion scale (*Icerya purchasi*) (From deBach, 1964; courtesy of Chapman and Hall).

combat the coconut moth *Levuana iridescens*; 32 750 parasitized larvae were released in Fiji, and control over the whole island was obtained in two years. No doubt success was due partly to the mild climate enabling generations of the coconut moth to overlap continuously so that caterpillars were always available for parasitization, but also to the fact that the island is fairly small and isolated.

Glasshouse whitefly in Britain

This example is included because it is one of the few early British examples. In 1926 a gardener in Hertfordshire noticed that his whitefly scales (*Trialeurodes vaporariorum*) were not the normal translucent white colour, but black. He sent these black scales to Cheshunt Research Station in Hertfordshire for identification, where the scales were found to have been parasitized by a small wasp, *Encarsia formosa*. Cheshunt bred and released the parasite to growers from 1929 until the 1940s, when the use of DDT ended the practice. However, the use of *Encarsia formosa* was resurrected in the 1970s as part of a biological control package in glasshouses (see Chapter 13, Early integrated control). The technique needs good temperature control in the glasshouse, because the parasite is rather sluggish at low temperatures and really too effective at high temperatures when it overtakes the whitefly and then dies out itself.

Cassava mealybug in Africa

This is one of the most recent examples of successful biological control. The mealybug (*Phenacoccus manihoti*) was accidentally introduced and spread to 30 countries in Africa. A search in the presumed area of origin of the mealybug, South America, led to the specific parasite *Epidinocarsis lopezi* being introduced to Nigeria in 1981. Since then it has been released at 30 sites, and by 1986 it was established in 13 countries. *Epidinocarsis lopezi* was chosen after studies showed that, although its reproduction is rather slow compared with its host, it has a special feature which makes it so effective. Although it cannot prevent population explosion of the pest when the latter is breeding rapidly, there is also a time when the pest is hardly breeding and restricted to relatively few small colonies. *Epidinocarsis lopezi* can locate these small colonies by an odour released by the plant at the point of damage, and at this time it succeeds in reducing the already small pest population to extremely low levels.

Biological control of vectors of malaria

As pointed out earlier, biological control of medical and veterinary pests and vectors is far less practical than for crop pests. Whereas we had several hundred agricultural successes from which to select just a few landmark examples, the medical examples are very limited and few can claim the level of success attained in agriculture. We could not have found any examples to compete

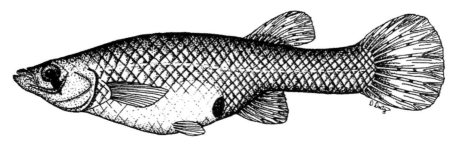

Fig. 7.9. *Gambusia affinis*, the 'mosquito fish' which has been introduced to many countries to control mosquitoes, especially anophelines (courtesy of D. A. Dritz).

with those above if we had combined the agricultural and medical examples before choosing a small selection; this is why we have selected malaria as the most important example in the medical area.

When rice fields have been stocked with 250–750 female *Gambusia affinis* (top minnow) (Fig. 7.9) per hectare significant reductions in mosquito larvae have sometimes been reported. For example these fish caused about 95% reduction in larvae of *Anopheles pulcherrimus* (a malaria vector) in rice fields in Afghanistan, although in California fish in rice fields have only given about 40% reduction of *Anopheles freeborni* larvae. In Indian towns *Gambusia* and *Poecilia reticulata* (guppy) have been routinely bred and introduced into wells, cisterns, water-tanks and other man-made reservoirs to control the urban malaria vector, *Anopheles stephensi*. In some wells they survived well and seem to have given 'effective', albeit localized, mosquito control, but many wells dried up and so fish had to be repeatedly introduced. In Greece in the 1930s *Gambusia* substantially reduced populations of *Anopheles sacharovi*, but their impact on malaria transmission remains dubious, and the disease was not eradicated until many years later – by spraying houses with DDT!

Fish have been used in parts of Iran, Afghanistan, Somalia, Ethiopia, Turkey and the Ukraine in attempts to control malaria. These are arid or semi-arid situations where water is scarce and mosquitoes are more or less confined to breeding in man-made reservoirs or tanks and it seems that in these situations fish can significantly reduce mosquito numbers and may cause a reduction in disease transmission.

In trials in Colombia in 1983 it was reported that, after the parasite *Romanomermis culicivorax* had been introduced into ponds, malaria prevalence was reduced from 21.7 to 0.81% over a period of 27 months. But then *R. culicivorax* no longer survived and malaria transmission rose to previous levels. Unfortunately it was concluded that there was little likelihood of *R. culi-civorax* being able to sustain control of either mosquitoes or the diseases they transmit.

Biological control in stored products

Rather as for malaria, biological control of stored product pests is best discussed in general, rather than trying to select a few examples, as was possible with crops in the field to illustrate different aspects of how biological control has been used successfully. At first sight, the enclosed and controlled conditions in which grain etc. is often stored would suggest similarities with glasshouses, where biological control has been so successful. However, consumer tolerance for the presence of pests in food is zero, and even a biological control agent is not an acceptable contaminant.

In spite of this zero tolerance for insects of any kind, there have been some fairly successful liberations of biocontrol agents against stored product pests. Probably the most widely released predator has been the histerid beetle *Teretriosoma nigrescens* against a major pest of stored grain, the bostrychid beetle *Prostephanus truncatus* (the larger grain borer).

Principal reasons for the failure of biological control

Climatic mismatching

This was a major problem in the early days of biological control, but climatic adaptation is now checked during the quarantine stage of a biological control programme.

A very good example of climatic adaptation is the biological control of the walnut aphid (*Chromaphis juglandicola*) in California. This aphid was thought to have originated from Europe, and in 1961 the parasitic wasp *Trioxys pallidus* was introduced from France. This parasitoid established successfully on the coastal plain of California but not in the main walnut growing area of the Central Valley. This is an irrigated desert area, with considerable temperature extremes between summer and winter. A second search for natural enemies therefore centred on the Central Plateau in Iran, which has a temperature cycle similar to that of the Central Valley of California. *Trioxys pallidus* was also found here, and obviously was a different ecotype of the species. The importations from Iran in 1968 were established successfully in the Central Valley.

Absence of flowers

As mentioned earlier, modern herbicides can eliminate flowering weeds in a crop. Although the need of most biological control agents for flowers is now recognized, some failures earlier in the history of biological control may have been caused by the absence of flowers. A series of articles by George Wolcott in the 1940s tells the story of repeated failures of a biological control programme in Puerto Rico due to the absence of suitable flowers for the adult hunting wasp *Larra americana* that were being released to prey on mole crickets (a species of *Gryllotalpa*) in cereal fields. In a later programme in Mauritius, Wolcott introduced the weed *Cordia* before beginning his biological control releases. Interestingly enough, *Cordia* became a major weed of Mauritius and was later controlled in the 1950s by a different entomologist through the introduction of a herbivorous beetle from Trinidad to Mauritius. Wolcott therefore appears to have been eventually successful in a biological control programme, though at second hand.

Absence of alternative prey

Population dynamics theory (Chapter 2) might suggest that density-dependent responses of the natural enemy should prevent the prey population being reduced to levels so low that the natural enemies risk starvation. However, such population dynamics do not take account of ecologically catastrophic events such as harvest, insecticides, soil cultivation and crop rotation. These are an integral part of agriculture and, like sudden climatic events such as storms, may cause a more rapid reduction in prey density (an example of density-independent, and often catastrophic, mortality) than the density-dependent responses of the natural enemy can compensate. Indicative of this is that, as mentioned earlier, biological control programmes have been easier to implement and sustain on perennial than in the more chancy environment of annual crops. Similarly many medical and veterinary pests often colonize unstable habitats, such as mosquitoes breeding in small and transient collections of water, and house flies and related flies breeding in dustbins and rubbish dumps.

The example cited earlier of the mulberry scale (*Pseudaulacaspis pentagona*) in Puerto Rico, which became a pest again after many years of control by a ladybird, suggests that a slight mismatch of predator voracity to pests' population increase rate may take some time to show that alternative prey were a necessary component to sustain the system.

Attack by predators and parasitoids

The multiplication of a newly introduced biological control agent presents local predators, parasitoids and diseases with a potential new food resource. Thus the biological control agent may itself suffer from its own biological control. When ladybirds were introduced to control scale insects in Mauritius, lizards found the beetles much to their liking and endangered the biological control programme. At this stage, the introduction of predatory birds to control the lizards was considered!

Another example of the attack on a biological control agent by its own natural enemies stems from the largely successful biological control of the Kenya mealybug (*Planococcus kenyae*) on coffee in Africa by the wasp *Anagrus* species. As we shall see this example also has another moral – that a change in crop management may effect some change which destroys an existing pest/natural enemy equilibrium. The multiple-stem system of growing coffee was introduced for improved yields, but on this new system the biological control of the mealybug began to fail. Another parasitoid wasp (*Pachyneuron* species) started attacking *Anagrus* on the multiple-stem plants. Such parasitoids of other parasitoids are called 'hyperparasitoids'. The attack on *Anagrus* in the new multiple-stem system happened because *Pachyneuron* flies at a low height and so encounters both the mealybug and *Anagrus*, which the multiple-stem technique brings near to the ground. The problem this caused to the biological control was sufficient for some growers to revert to the older and less productive single-stem system in some areas.

Is biological control natural?

Biological control uses natural organisms that have a long history of evolution, most of it long before the advent of agriculture. We can therefore assume that the carnivores alive today have succeeded in keeping their food supply safely well above rarity level over this long period. The co-evolutionary forces on both prey and natural enemy will have led to neither species being endangered by the carnivorous activity of one of them. Sometimes natural biological control works sufficiently well that insects do not reach population levels that would cause a grower concern. Most examples of successful biological control involve such pest and natural enemy combinations. The pest has been imported to a new country or continent without its natural enemy, or the use of insecticides or other interferences has led to the herbivorous species becoming a pest problem. Biological control then involves restoring what has been the natural balance in the country where pest and natural enemy have evolved together. It may also be possible (see Chapter 13) to enhance the effect of the natural enemy in various ways. It is also possible (again see Chapter 13) that for various reasons biological control cannot be transferred so easily from where it is working into a new environment.

Many herbivorous insects on crops never become pest problems; the fact that natural biological control may often be responsible has rarely been investigated. However, in many other situations natural biological control is inadequate to keep an insect below pest status. For example, in some African countries and at certain times of the year adult biting densities of anopheline vectors must be reduced 1000-fold before malaria transmission is interrupted. This is a virtually impossible goal for biological control. Similarly, in Kenyan rice fields, although naturally occurring aquatic predators cause about 90% mortality of the immature stages of *Anopheles arabiensis*, the major malaria vector in the area, the numbers of adults emerging are sufficient to maintain high levels of malaria transmission. One approach would be to augment the population of natural enemies, but it is usually difficult to achieve any significantly increased mortality by increasing populations of predators and parasites that occur naturally, and augmentation would normally involve releasing exotic agents.

Such augmentation is a 'trick' to get the system working unnaturally, although we must then expect the system to evolve counter-adaptations to the trick we have introduced. Importing additional natural enemies may not be as effective as attempting to trick the existing system in other ways, since competition between the indigenous and imported enemies may work against an improved biological control outcome. Examples of the kinds of tricks that may be employed can be found above in the description of the techniques of biological control – two examples are the removal of the agent's diseases during quarantine and the use of ecotypes of the agent with lower temperature development thresholds. As also mentioned above, there is increasing interest in adding alternative prey, flowers and other requirements of biological agents to the habitat, and manipulating the agents with behaviour-modifying chemicals.

Another unnatural intervention for the future is genetic modification of biological control agents. Although the techniques for genetically modifying animals are still rather difficult compared with those for plants and bacteria, such modification is already possible, and will become more routine with time. Disarmed animal viruses (i.e. viruses with the gene causing pathogenicity replaced with the desirable gene) can be used to infect the target animal and transfer the gene, or the desired DNA can be directly injected into insect embryos. Although the imagination can probably conjure up many desirable attributes of natural enemies that might be transferred in this way, the obvious one is surely resistance to insecticides. After all, the genes for this are already present and well-known in other animals, namely the pests the biological control agents are there to control. The transfer of such genes to predators, for example, would remove one of the important limitations of biological control that they, like their pest prey, are killed by pesticides (see earlier). It seems a nice concept that the pests could be donors to their own biological control agents of genes for survival.

8

Insect pathogens

Introduction

Microbial pesticides are an increasingly important area of biological control. The current *Biopesticides Manual* (Copping, 2001) lists 44 commercially available products used for the control of insect pests. Such 'germ warfare against insects' has quite a long history. The Russians carried out the first experiments as early as 1884, and a commercial preparation of a fungus (*Beauveria*) was available in Paris in 1891. Like insecticides, pathogens can be stored for a period, marketed in drums, diluted with water and passed through a spraying machine.

Pathogens are, of course, biological control agents. However, they have several important differences from animal natural enemies, particularly much shorter generation times and vast production of propagules; also that they can be sprayed from traditional sprayers. They do not seek out their prey, but rely on chance contact. Moreover, in some cases an insect toxin produced by the pathogen is used rather than the living organism, and then there is no multiplication or recycling of the pathogen. For all these reasons, it seems appropriate to deal with pathogens in a separate chapter. Although a great variety of pathogens, viruses, bacteria, protozoa and fungi have been recorded from medical and veterinary pests, very few have actually shown promise as control agents. There seem to be several reasons for this. It would prove logistically difficult, if not impossible, to spray the numerous and scattered resting sites of dipterous vectors, such as mosquitoes and simuliid black flies, with pathogens; consequently control would have to be aimed at their aquatic larvae (but see p. 183 on how entomopathogenic fungi can kill ticks). A problem with applying pathogens to aquatic stages of pests is that many potential pathogens, such as *Coelomomyces* species (fungi), have very complex life histories, and many other pathogens do not disperse well in water and so infection rates are very low. Nevertheless good control of mosquito and black fly larvae is achieved with the bacterium *Bacillus thuringiensis* var. *israelenis*, although it is applied as a microbial insecticide (see p. 188). The situation regarding the use of pathogens against crop pests is much more promising.

Advantages of pathogens

As a form of pest control, pathogens have some quite striking advantages:

(a) *Target specificity.* In contrast to insecticides, pathogens are very selective. They often kill one or a limited number of orders of insects, and some are even specific within an order. They can therefore often be applied with complete safety for natural enemies.

(b) *No toxic residues.* Pathogens have a limited life outside their host and therefore very short persistence in the environment. Also insect pathogens, by virtue of the specificity just mentioned, are generally non-toxic to humans and domestic or farm animals.

(c) *Resistance development is unlikely or at least slow.* As far as we use naturally occurring living pathogens, resistance to their developing in the pest is an unlikely event. Pests and pathogens have already been associated for centuries, and it is likely that a succession of resistance mechanisms evolved by the pest in the past has been countered by the pathogen, and that any future mechanisms would be overcome in a similar way. However, this might not be as inevitable with pathogens changed by genetic engineering, or when the pathogen has been replaced purely by its toxin (e.g. as with the bacterium *Bacillus thuringiensis*). In fact resistance in mosquitoes and Lepidoptera to this insecticide (in spite of its natural origin) has already occurred.

(d) They are *usually compatible with pesticides.* In contrast to biological control, many pathogens are compatible with insecticidal usage and can often be used in combination with them.

Being specific and environmentally friendly with little danger of development of resistance is surely a job description for products ideal for use in pest management programmes. And there are further applications for pathogens. They can be brought in when traditional insecticides have been rendered ineffective by resistance in the pest, such as in West African simuliid vectors of river blindness and diamond-back moth (*Plutella xylostella*) in South-East Asia. Unfortunately, in the latter example, use of the toxin alone has already selected tolerant populations of the pest in several locations. Pathogens also have a clear role where the application of pesticides is restricted, for example close to harvest or on environmental or cost grounds. Thus they have been found extremely useful in some low-input cropping situations such as pasture and forestry.

(e) *Ease of genetic modification.* The genetic modification of insect pathogens is technically straightforward and can effect great improvements in pathogenicity. Research is an active area, and products have been marketed. The prospects for genetically modified pathogens are good, though it is possible that the higher selection pressure they put on the pest may accelerate the appearance of tolerant pest strains.

Disadvantages of pathogens

Unfortunately, pathogens also have serious disadvantages, particularly in the prospects of successful commercialization. However potentially ideal they are for pest management, for commercial reasons it is unlikely that they will fulfil this potential, at least for some time to come. What a pity! Among the reasons for this are:

(a) *Specificity can be too high.* The high specificity of pathogens sets economic limitations, since specificity also means market restriction. Development costs for pathogens may be less than for insecticides, but are still sufficiently high to require larger markets than a restricted range of target pests can provide.

(b) *Problems of shelf-life and persistence in the field.* Pathogens are living organisms, often with a very short life-span in nature, and so it can be extremely difficult to produce them on a factory scale and store them while retaining their virulence. Many will also need to be formulated with additives such as wetters and spreaders. Storage becomes a particular problem once the product has left the manufacturer; the organisms may need to be shipped and then stored by merchant and farmer, often under less than ideal temperature and humidity conditions (e.g. in the tropics). Once applied in the field, they may fail if conditions are too dry or too hot, or if the pH conditions in the water, soil or on the leaf surface, are outside certain limits. Many pathogens are also very sensitive to ultraviolet radiation, and so are rapidly destroyed by sunlight. Thus chemical radiation 'shields' are a frequent additional component of the formulation. Once applied, UV radiation is not the only problem. Many pathogens, especially the fungi, require adequate if not high humidity if they are to survive for any length of time. There is also evidence now that the field efficacy of viruses may be dramatically reduced on certain crops (especially cotton), apparently because of lethal exudates on the surface of these plants.

(c) *A threshold population may be required.* If we hope to introduce a disease, and then for it to spread through the pest population, ideally we need to have reasonably high pest populations. There are therefore critical threshold populations, often above those acceptable to the grower, below which the disease will not spread. Largely because of this limitation as well as the problems of appropriate environmental conditions such as humidity, pathogens (except for applications to the soil or water) are rarely used other than as 'microbial pesticides' – i.e. they are applied to control the pests with the initial application (inundative use).

(d) *Unsightly insect corpses.* Although pathogens leave no toxic residues, they do leave the corpses of their victims. Because the pathogen has evolved to infect other individuals, it has usually left the corpse adhering firmly to the plant. This may force the grower into expensive washing procedures before he can sell his produce. This is not a constraint with medical pests and vectors such as mosquitoes and simuliid black flies, whose dead larvae are left behind, unseen, in the water. Nor are corpses on the foliage a problem in forestry.

(e) *Problems of development.* Mention has already been made of the commercial difficulties caused by selectivity and viability through production and storage. Also, development research is not as straightforward as with insecticides, since the direct toxicity which can be demonstrated in the laboratory is perhaps less important than certain behavioural and biological properties of the host insect in the field. These ultimately determine the contact between the pest and a sprayed pathogen. Pathogens are developed by greatly increasing diseases first located in natural insect populations, and this may often mean expensive rearing of large numbers of pests to multiply the disease. Consequently a major constraint faced by potential candidates is the necessity for *in vivo* production which may make mass production impossible, or very difficult. However, fungi and bacteria are amenable to multiplication by the modern techniques of biotechnology, and these pathogens have received most attention in recent years.

Finally, registration procedures still need considerable further modification from the very expensive data protocols required for toxins

Fig. 8.1.   A pea aphid (*Acyrthosiphon pisum*) killed by the fungus *Erynia neoaphidis*, showing the distance over which fungal spores from the cadaver are projected (courtesy of J. K. Pell).

such as insecticides, for pathogens to become attractive for commercial development.

Types of pathogens used in pest control

Contact microbials: fungi

Invasion of insects by fungi results from a spore landing on the cuticle of the insect, and the germ tube then penetrating the cuticle directly. The short distance spores are projected from an insect corpse (Fig. 8.1) means that only nearby insects can be infected. Early work focused on the genus *Beauveria*, especially *B. bassiana*, used particularly against the cabbage looper (*Trichoplusia ni*). Pest control with sprays of fungal spores has always been potentially unreliable, since the spores require moisture to sporulate and really quite high humidities are required for success. An obvious use of fungi is therefore against soil pests, and the fungus *Metarhizium anisopliae* has shown considerable promise for soil application. Fungi have also been used in glasshouses,

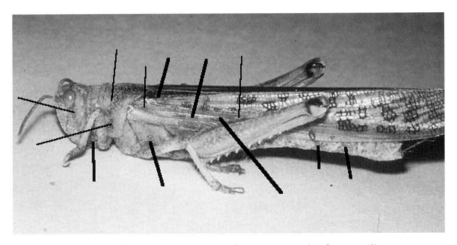

Fig. 8.2. A locust showing symptoms of 'green muscle' fungus disease (Courtesy of R. P. Bateman). Thin lines have been added to point to areas where the cuticle shows a characteristic reddening, and thick lines areas to where the green fungus appears externally.

where high humidities are not so difficult to achieve and maintain. Much research has been done on a fungus, *Verticillium lecanii*, originally found in scale insects (Coccoidea). Different isolates of this fungus have proved effective against aphids and whiteflies (Aleyrodoidea), and different formulations suitable for the two purposes have been marketed.

However, research seeking to use green muscle disease (*Metarhizium anisopliae*) for control of locusts (Acrididae) (Fig. 8.2) in arid climates has broken through the dependence of fungi on high humidities by exploiting the application of low volumes of spray per hectare made possible with spinning cup sprayers (see p. 89). Ultra-low-volume application has made it possible to apply the fungus with oil and not water as the carrier. The oil does not evaporate, and protects the spore from desiccation. Of course the spore still needs water for germination, but it can obtain this from transpiration through the cuticle of the insect.

Fungi are compatible with many herbicides and insecticides. In the locust example above, a very low dose of pyrethroid insecticide applied at the same time may prove an important contribution to successful control. The fungal disease takes several days to inhibit locust feeding, during which time the locusts can do a great deal of damage. Experiments have shown that this window of damage can be almost completely closed with the sublethal insecticide dose; not only does feeding cease almost immediately, but the eventual inhibition by the fungus is also advanced by a couple of days. In terms of compatibility of insect-pathogenic fungi with pesticides, however, caution is needed where fungicides to control plant diseases also have to be applied to the crop.

The oil formulation technique has opened up a new possibility for the use of fungi. Their safety to humans makes them attractive for use in stored foods, where insecticide residues are clearly undesirable. However, before the breakthrough with oil formulation, the dry conditions necessary for food (especially grain) storage made the use of fungi out of the question.

In contrast to the agricultural scene there are currently no fungi that can be recommended for effective control of medically important pests, despite considerable interest and research in the 1960s to 1980s, often sponsored by the WHO, into fungal pathogens such as *Coelomomyces, Langenidium* and *Culicinomyces*. With ticks, however, there may be hope for control with fungi. Experiments in several countries have shown that, when *Beauveria bassiana* or *Metarhizium anisopliae* are sprayed on cattle or on vegetation harbouring ticks, there is high mortality of ticks in several genera. More research is needed, however, before commercialization can be contemplated.

Ingested microbials

These pathogens rely on ingestion by their host to initiate an infection. They are therefore adapted to having a fairly resistant stage in the life cycle which enables their survival on relatively dry surfaces, such as a leaf, until ingestion by the pest. Such pathogens are therefore rather less humidity dependent than the fungi, though the capacity for survival of the resistant stage is still short in comparison with the residual effective life of most insecticides.

Viruses

More than 300 viruses have been isolated from about 250 important pest insect species, yet less than ten are in commercial production. Although viruses have the disadvantage that they have to be cultured in living insects, there have been some commercial products. The most successful pest control viruses are the Baculoviruses in the Baculoviridae, a family of viruses known exclusively from arthropods. The subgroups nuclear polyhedrosis viruses (NPVs) and granulosis viruses (GVs) contain the most important viruses used in pest control (Fig. 8.3). A major success with viruses has been the control of the sawfly, *Neodiprion sertifer,* on forest trees in Canada using a nuclear polyhedrosis virus

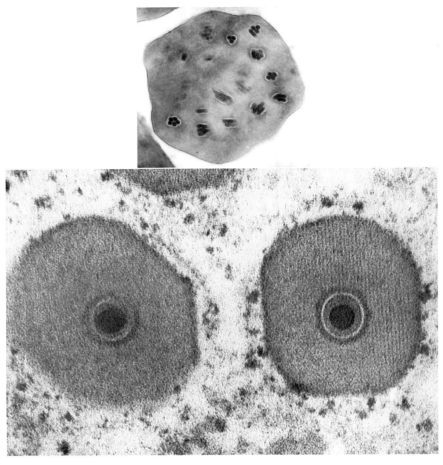

Fig. 8.3. Electron photomicrograph of a polyhedron of a nuclear polyhedrosis virus of the celery looper, *Anagrapha falcifersa*, showing the enveloped virions (above; courtesy of D. Hoffmann and D. Hostetter) and granulosis virus of codling moth, *Cydia pomonella* (below; courtesy of D. Winstanley and HRI).

(Fig. 8.4). There is currently interest in controlling the armyworm *Spodoptera* (Noctuidae) on wild plants in Africa to keep populations low and prevent its mass migration to crops following a build-up of numbers in the grassland areas where it breeds. In this instance the disease inoculum is being obtained from natural infestations. Viruses can be applied at a very low dosage; for example, in work with naturally diseased caterpillars on cabbages, the disease of less than two infected larvae was applied per hectare.

Fig. 8.4. Larvae of the European sawfly (*Neodiprion sertifer*) dying from infection with a nuclear polyhedrosis virus (courtesy C. F. Rivers).

Although nuclear polyhedrosis and cytoplasmic polyhedrosis viruses occur in several medically important insects, there is very little prospect of their being used in their control. Why? Because apart from having to be grown in living insects or cultured in insect cell lines they have not proved very infectious, and mortality rates are usually below 25%.

The development of commercial virus preparations for agricultural usage, and their acceptance by the registration authorities, has been somewhat hampered by fears as to the safety of distributing them in the field in case they then mutate, to attack humans (another 'AIDS'?) or domestic animals. However, many virus diseases of insects are specific to their hosts and already occur naturally in host populations. Their use would need to be very widespread before the quantities of pathogen sprayed by man outweighed the inoculum available in natural populations. However, it could also be argued that the development of viruses by man for pest control involves the selection, as well as the production by genetic modification, of new strains of particular potency for particular target organisms and that the release of these in the field is not equivalent merely to an increase of the natural inoculum. Viruses are easily spread in the field on the feet of birds, and so can be transported well outside the target area. Also, although a particular virus may attack only Lepidoptera,

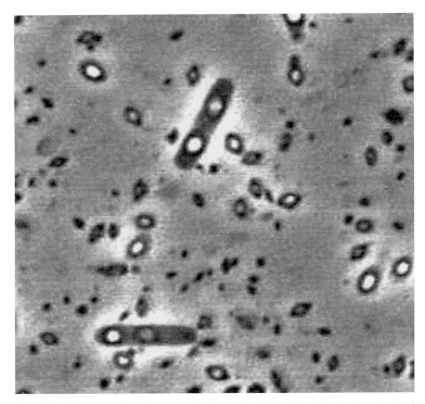

Fig. 8.5. Phase contrast microphotograph of *Bacillus thuringiensis*. In the elongate parasporal bodies, the lighter circular structure is the bacterial spore while the darker one is the endotoxin crystal (courtesy of J. Deacon, University of Edinburgh).

the host range within the order may be quite unpredictable. These two aspects mean that there is a risk that virus may be carried to non-target pests, including into conservation areas. There is thus an advantage in preventing the virus spreading by using it merely to take the real cause of death, a rapidly acting poison, into the insect. An example is the genetic engineering of a Baculovirus to express snake venom.

Bacteria
Bacterial preparations usually formulated as powders typically contain about a thousand resistant parasporal bodies per milligram. The powders are wetted and sprayed and the pest may then ingest vegetative bodies while feeding. Each parasporal body (Fig. 8.5) contains two structures, a spore and a protein

crystal (the endotoxin). Although the spore often releases some toxins, the important insecticidal element is the protein of the crystal. When the parasporal body reaches the alkaline conditions of the insect gut, the protein crystal dissolves.

The exact pH of the gut is critical and accounts for the considerable selectivity of action of the bacteria. Different genotypes are needed to kill Lepidoptera and Coleoptera, for example, and as yet no effective bacterium against locusts has been found. The pH sensitivity also accounts for the safety of these bacteria for humans. The mode of action differs between species of insect. Some are directly sensitive to the endotoxin, but more usually the endotoxin does not kill directly, but binds to the midgut wall leading to the formation of pores. These pores enable the toxic bacteria which are always present in the gut of animals to invade and kill through septicaemia. The bacterial spore itself does not germinate and propagate vegetatively until pH conditions change when the gut ruptures and the insect becomes a corpse. As the cadaver dries out, the bacterium resporulates to form new parasporal bodies. The bacteria have the commercial advantage that they can be 'brewed up' in huge numbers in nutrient solutions without the need for insect hosts.

Today there is considerable interest in developing the toxic protein carried by parasporal bodies as an insecticide in its own right, and the bacterial toxin most commercialized is that of *Bacillus thuringiensis*, often simply known as '*B.t.*'. A fresh commercial preparation of *B. thuringiensis* toxin has often compared favourably with normal insecticides for the control of insects. It has the great advantage of being harmless to honey bees and suitable for application close to harvest of edible foodstuffs and to aquatic habitats. Because it is so non-toxic to humans, the gene for expression of the *B.t.* toxin was used for the first generation of genetically modified pest-resistant plants. Whether this represents using the toxin as an insecticide and replacing the delivery system from a spraying machine with a plant, or whether it is a new approach to host-plant resistance is debatable, but in this volume such genetically modified plants will be discussed in more detail under the latter heading. It is increasingly apparent that different strains of *B.t.* can be isolated or genetically engineered to be more virulent and highly target-specific to different pest species. Thus although *B.t.* was first discovered in 1911 and was already available as a commercial 'insecticide' in 1938, none of the isolates was effective against medical or veterinary pests until 1976, when a new isolate was discovered in pools in the Negev desert, Israel. This strain has proved to be effective against larvae of mosquitoes, simuliid black flies, house flies and a few other Diptera. It was named *B.t.* var. *israelensis* (*B.t.i.*), or serotype H-14 as the WHO prefers to call it.

As early as 1980, nearly 30 'varieties' of *B.t.* were known, and the primary cause of the variable spectrum of activity of the varieties seems to lie in the spectrum of activity of the protein crystal toxin. The exceptional efficacy of *B.t.i.* against mosquitoes and black flies led to commercial production in 1981, and there are now several companies in North America, Europe and elsewhere making flowable concentrates, wettable powders, granules and slow-release briquettes of this microbial insecticide. In the USA, *B.t.i.* has become one of the most commonly used larvicides for controlling mosquitoes. However, the greatest success is in application to rivers and streams to kill larvae of simuliid black flies that have become resistant to temephos, the main insecticide used in the massive West African Onchocerciasis Control Programme (more details of this programme are given in Chapter 4) to kill the vectors of river blindness (onchocerciasis). Although this solves the resistance problem, larger quantities of this microbial insecticide have to be dropped from helicopters for effective control compared with temephos, and the insecticide does not carry downstream so far as does temephos, necessitating more focal drop points.

Of course, as mentioned earlier, using the toxin in isolation from the living bacterium is likely to result in resistance in the pest just as to any other insecticide, and increasing the virulence by selection or genetic engineering will only hasten this event. It is therefore not surprising that resistance to the *B.t.* toxin was reported in *Plodia interpunctella*, a lepidopteran grain pest, in 1985 and (as mentioned earlier) in the diamond-back moth *(Plutella xylostella)* as early as 1990, and is now quite widespread. Although resistance to both *B.t.i.* and *B. sphaericus* (see below) has been reported in a few mosquito species, presently this does not pose operational problems but should serve as a warning that there may be control difficulties in the future.

Two other bacteria used in pest control are *Bacillus popilliae*, which is available specifically against beetles, especially the Japanese beetle, *Popillia japonica*, and is mixed into the soil to combat this underground pest of grass in the USA, and *B. sphaericus* which is used to kill mosquito larvae. Unlike *B.t.i.*, *B. sphaericus* recycles in some situations and may persist for some weeks or months.

As long ago as 1956, the Chinese attempted to control codling moth with a bacterium, by spraying bacteria-infected eelworms (nematodes) to enter the larval canal into the fruit so as to reach the larvae. Nematodes ingested by the codling caterpillars then pierced the gut to allow lethal access for the bacterium. More recently, use of nematodes to carry bacterial toxin to pests has led to new commercial products (see Chapter 7).

Microsporidia

Until relatively recently, microsporidia were included in the phylum Protozoa. However, most experts now regard them as a separate phylum, with no known affinity to other acellular organisms. Microsporidians in the genus *Nosema* have long been investigated as insect pathogens for biological control, and *N. locustae* has been used successfully against grasshoppers in rangeland in the western USA. Although field tests with *Brachiola (= Nosema) algerae* and *Vavraia culicis* have been conducted against mosquitoes there is really no hope that microsporidia will become a realistic control tool for these problems. A major problem with microsporidia is that they can only be propagated on living insect hosts, which makes their commercial multiplication extremely expensive, and also that they are more likely to debilitate the insect than to achieve a rapid kill.

Conclusions

In summary therefore, pathogens have huge advantages as safe, environmentally friendly and selective tools for pest control. Many of the practical problems of using them in the field have been overcome, but it is the commercial limitations of expensive registration, high selectivity and storage and field viability which remain hurdles to their availability.

9

Genetic control

Introduction

Genetic control of pests and vectors is often called autocidal control, except where sterilizing chemicals (chemosterilants, see later) are applied to field populations. Autocidal control is particularly ecologically friendly. This is because, unlike biological control even, nothing alien is introduced into the environment, albeit that the target organism is genetically modified in some way. However, considerable ecological knowledge of the population dynamics of the target pest is usually required, especially as regards mating behaviour and dispersal. Unlike most control strategies the emphasis is not so much on death processes as on reducing birth rates. The target pest must be capable of being cultured and reared in usually enormous numbers, which immediately limits the species that can be candidates for genetic control.

Several techniques have been proposed for genetic control (e.g. hybrid sterility, cytoplasmic incompatibility, chromosomal translocations), but the most widely used and most successful has been the sterile-insect technique (SIT) (sometimes called sterile male technique (SMT) or sterile insect release method (SIRM)). This usually involves the mass release of sterile adult males. The technique is particularly suited for insect species that mate only once or, if there is multiple mating, sperm from the first mating is the sperm that fuses with the eggs. Purists would argue that SIT is not true genetic control because the sterility induced is not inherited, nevertheless it is generally considered a form of genetic control.

Brief descriptions are presented of some of the genetic methods that have been proposed for the control of pests and vectors. Not all have proved successful and some have not yet been field-tested, but the examples given illustrate some of the limitations of the techniques and the enormous gap that has to be bridged between academic laboratory-based experiments and successful field implementation.

Sterile-insect release technique

Sterility is produced by applying radiation or chemosterilants to the immature stages during mass laboratory rearing. The application of chemosterilants directly to field populations is described later. The idea is that, following mass release of sterilized male adults, they should compete with fertile native males in seeking out females so that a high proportion of the native females are mated by sterile males. The accompanying release of sterilized female insects, alongside infertile males, is usually inevitable with the method, but normally has no impact on the population of the species. However, if the insects are disease vectors and large numbers are released they can cause an increase in disease transmission. This would be an unacceptable situation. Consequently, with insects like mosquitoes there should be some method of sex separation to minimize the release of females.

The control of the New World screwworm fly (*Cochliomyia hominivorax*), a serious economic pest of cattle, is a landmark in the history of pest control. Edward Knipling considered the screwworm an ideal target, because it fulfilled several theoretical requirements he believed essential for success of the technique, namely:

- A method for mass rearing of male flies.
- The released males must disperse rapidly through the native population.
- Sterilization must not affect sexual competitiveness.
- Preferably females only mate once.

Knipling (1955) also developed a model (Table 9.1) on how a pest population might decline in subsequent generations with a constant number of sterile male flies released per generation. Clearly the natural population has to be swamped with sterile males, so Knipling suggested that sterile males should be released only after the native population was reduced by insecticides or was naturally low because of climate or season.

Weekly releases of 136 000 New World screwworm flies for nine weeks in the 1950s sterilized by gamma radiation (using a 5000 Röntgen cobalt bomb) resulted in the successful eradication of the screwworm from the island of Curaçao. Later the fly was eradicated from south-eastern USA and good

Table 9.1. *Theoretical population decline in each subsequent generation when a constant number of sterile males are released among a natural population of 1 million females and 1 million males. From Knipling (1955).*

Generation	Number of virgin females in the area	Number of sterile males released each generation	Ratio of sterile to fertile males competing for each virgin female	Percentage of females mated to sterile males	Theoretical population of fertile females each subsequent generation
F_1	1 000 000	2 000 000	2:1	66.7	333 333
F_2	333 333	2 000 000	6:1	85.7	47 619
F_3	47 619	2 000 000	42:1	97.7	1 107
F_4	1 107	2 000 000	1807:1	99.95	Less than 1

suppression was obtained in south-western USA. These were the first successful large-scale control operations using the sterile male technique. Since then the screwworm has been eradicated from south-western USA, Mexico and from most Central American countries, using large-scale production facilities (Fig. 9.1) for sterile insects located on a disused airfield site not far from the Mexican border. At the height of the campaign, 50 tons of blood and meat per week were converted into 150 million sterilized and released flies. The releases were made along the USA–Mexico border, in a strip 3000 km long by 400 km wide. However, 90 000 cases of screwworm attack were reported in 1972 from the reportedly cleared area. The continual inbreeding in the laboratory had weakened the culture to the point where Knipling's condition, that females would mate readily with sterile males, was failing. Since this event, it has been realized that the rearing stock needs to be periodically revitalized by the introduction of wild insects.

Probably through transport of infected animals from the New World, the screwworm fly appeared in the Old World (Libya) at the start of the 1990s, with 3000 new cases a month by September 1990. Because importing US expertise to Libya was politically incorrect, sterile insect release was instead organized in the name of the Food and Agriculture Organization (FAO). Releases started in December 1990, and the last case of attack by screwworm was noted as soon as April 1991. Releases could cease that October. It was inevitable, with the publicity following the successful SIR programme in the

Fig. 9.1. The factory used to produce sterile male screwworm flies in the south-eastern United States (courtesy of The National Agricultural Library, Beltsville, Maryland).

south-eastern USA, that the method would be tried with many other targets, both in the USA and elsewhere. In Europe, the International Atomic Energy Agency in Vienna has for many years had a team of entomologists working on SIR programmes.

The sterile-male release method has been applied in attempts to control *Aedes, Culex* and *Anopheles* mosquitoes in countries such as India, El Salvador and the USA. Unfortunately none of the field trials in the 1960s in India and the USA with radiation-sterilized male mosquitoes achieved accepted levels of population reduction. Mostly failures were due to poor fitness of released males, caused either by long periods of colonization or by large doses of radiation. Later the daily release over eight weeks of 141 400 sterilized male *Culex quinquefasciatus* effectively controlled this species on Sea Horse Key, a very small island off the Florida coast. However, field trials near Delhi, India, against the same species failed, mainly because the released sterilized males were not as competitive as native males in finding mates, and because of immigration of fertile females from surrounding areas.

In 1971 in a very small area at Lake Apastepeque, El Salvador, an estimated 4.36 million chemosterilized male *Anopheles albimanus* were released over five months. The larval population of this important malaria vector, that had become resistant to a wide range of insecticides (mainly through cocktails of insecticides being sprayed on the cotton crop), was reduced by more than 99.9% – a marvellous achievement. Then in 1976 a larger trial covering 20 km^2 near the coast involving the daily release of one million sterilized males a day for four months was undertaken. In combination with methoprene (an insect growth regulator) as a larvicide, a considerable reduction in population size was achieved, although migration of fertile females into the area prevented eradication. This is generally regarded as a successful experiment, at least academically. But even if eradication over the trial area could have been obtained, how could this success be extrapolated to larger areas when such vast numbers of *A. albimanus* had to be reared and released for just an area of 20 km^2? This trial is noted for the elegant manner in which female mosquitoes were prevented from being released. Sex-linked resistance to propoxur was obtained, enabling eggs destined to become females to be killed; thus colonization produced just male mosquitoes.

The low reproduction rate of tsetse flies theoretically makes them good candidates for genetic control. An initial trial in the 1970s involved aerial insecticidal spraying of a small island in Lake Kariba, Zimbabwe, followed by release of puparia of *Glossina morsitans* sterilized by a chemosterilant. About 98% population reduction was achieved after nine months. Then in the early 1980s, some 650 000 sterile males were released along 32 km of riverine vegetation against a relatively low density population of *Glossina palpalis* artificially isolated from adjoining populations. The fly was partially eradicated. In Tanzania, after two aerial sprayings against a high population density of *G. morsitans*, some 351 000 sterile male flies were released during a 13-month period over a 200-km^2 area and 90% control was achieved.

Glossina austeni, a vector of animal trypanosomiasis to cattle, was eradicated in 1997 from the island of Unguja by sequential releases from light aircraft of more than 8.5 million gamma-sterilized male flies over a period of about 40 months, but only after insecticidal treatment of cattle had reduced the population. Success was possible because the island was only 1650 km^2 in area and was 35 km from the Tanzanian coast; hence the tsetse population was relatively small and isolated. Nevertheless the initial number of sterile males released had to be 10 times greater than native males to achieve a population decline, this ratio being much greater than the theoretical 3:1 ratio proposed by Edward Knipling for tsetse flies. It was also discovered that sterile males were not as competitive for mates as native fertile males.

Despite the successful eradication of *G. austeni* from Unguja island the reduced competitiveness of released sterile male flies, their high costs of production, need for prior insecticidal spraying if tsetse fly populations are at high densities, and the fact that trypanosomiasis is often transmitted in the same place by more than one *Glossina* species, usually makes this approach unrealistic.

As far as pests of crops are concerned, the many attempts there have been with SIR have led to hardly any successes. This is largely because the screw-worm fly is unusual in that the females accept sterile males, whereas females of most insect species seem able to recognize low fitness in a prospective mate. Also, most crop pest females mate more than once, so that a mating with a sterile male does not necessarily prevent their reproduction. SIR, however, has become one of the pest management options for control of the boll weevil (*Anthonomus grandis*), even though females of this weevil do mate more than once. The main problem is that younger irradiated males live longer than those irradiated later, but the older males are more competitive because they transfer more sperm. If the effective life of released younger weevils could be doubled, it would result in significant savings in what is a costly method.

It is the Mediterranean fruit fly (*Ceratitis capitata*) which has probably been the prime crop pest target for SIR, since it does mate only once. Programmes have been undertaken in North, Central and South America as well as in Egypt. Such programmes have usually begun with a reduction of the local population with insecticide, but even then success has been little better than partial. A recurring problem has been the lack of isolation of the population in treated areas, and that the factory-reared sterile males often mate several weeks earlier than is acceptable to the wild population.

One has to conclude that SIR really has to be limited to extremely harmful insects which also have the biological characteristics favouring success of the method. The high initial cost can only be recouped without additional recurrent cost once the pest has been eliminated. The economics of the method would be much improved if production could be limited to males. With some pests, including fruit flies, females do damage with attempted oviposition punctures and adult feeding, whether sterile or not, and female vectors can still bite and spread infections to man and livestock (see earlier). Moreover, the release of females is anyway undesirable, as the whole aim is to release many males among the minimum number of females. If the female gender could be genetically linked to a deleterious trait such as sensitivity to high temperature or a chemical (as described for *Anopheles albimanus* and propoxur above), exposure of the sterilized insects to the lethal factor would eliminate all females before release. Genetic manipulation (see the end of this chapter) offers such possibilities.

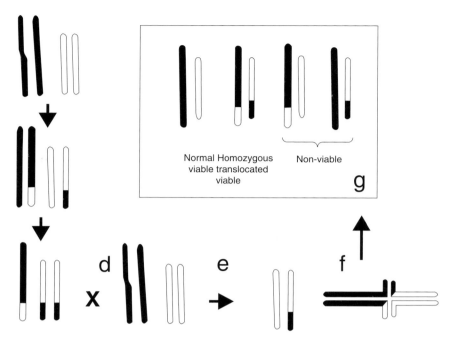

Fig. 9.2. Radiation-induced chromosomal translocations. (a) normal chromosomes; (b) chromosomes with induced translocation; (c) translocated homozygote crossing with (d) normal chromosomes gives (e) translocated heterozygote. At meiosis an aberrant pairing of chromosomes (f) then occurs, giving rise to four gamete possibilities (g).

Chromosomal translocations

Although the idea that inherited sterility caused by reciprocal chromosomal translocations might be used to control insects was originally proposed in 1940, it was some 30 years later before the possibility was seriously considered as a control mechanism. The genetics of this technique can be quite complex, but basically a simple chromosomal translocation involves the breakage of two non-homologous chromosomes and reattachment of the broken pieces to the wrong chromosomes, i.e. they have been translocated (Fig. 9.2). Such breakage (which may also occur naturally) can be induced by irradiation. At meiosis the translocated heterozygote produces different types of gametes, 50% of which are typed as 'unbalanced' and lethal. Consequently mating between a translocated heterozygote and a normal individual results in just 50% fertility.

Moreover, half the offspring surviving such crosses inherit the same translocation which is then passed to their progeny. But for field releases translocated homozygous individuals (Fig. 9.2) are more efficient in causing population suppression. This is because when they mate with normal individuals, the complete viable progeny consist of translocated heterozygotes which are semi-sterile. Thus only half the numbers of translocated homozygous individuals are needed to achieve the same degree of sterility in a population. Mating of translocated homozygotes can result in multiple translocations and production of even more genetically aberrant individuals.

The first partially successful field trial using translocated insects was in France in the 1970s and involved the mosquito *Culex pipiens*. Other trials have been conducted with *Aedes aegypti* in India and Kenya, and *Culex tritaeniorhynchus* in Pakistan, with very variable success. In the 1970s entomologists and geneticists produced translocations in tsetse flies (*Glossina austeni*), house flies (*Musca domestica*) and blow flies (*Lucilia sericata*), but there have been no field trials to determine whether such insects can give control.

No thought seems to have been given to researching the translocation approach for the control of pests of agriculture. There seems no obvious reason for this, other than the failure of most agricultural entomologists to read the medical entomology literature!

Hybrid sterility

The *Anopheles gambiae* complex consists of seven sibling species, of which *A. gambiae* s.str. is probably the world's most efficient malaria vector. In the laboratory these sibling species often mate with each other, but when any two species are crossed the resultant males are infertile, so if sterile hybrids were released into a population this should result in control. The idea was tested by first crossing in the laboratory *A. arabiensis* from Nigeria with *A. melas* from Liberia and then releasing some 300 000 hybrid pupae over two months to try and control a natural population of *A. gambiae* s.str. in Burkina Faso, West Africa. There was no control. This is not surprising when one considers that sterile hybrid males produced by crossing two species were supposed to mate with a third species, namely *A. gambiae*! What is surprising is the belief that this approach might have worked.

In laboratory trials male *Aedes albopictus* will mate with *Aedes polynesiensis* females, an important vector of filariasis, but the eggs are infertile. This led to an experimental release of *A. albopictus* on a small Pacific atoll of the Society islands to control *A. polynesiensis*, but the results were inconclusive and the introduced *A. albopictus* disappeared after four years.

Different genetic populations of the same insect species are not always compatible; this is especially true in the Lepidoptera. This order of insects is unusual in that the males (if this gender is defined by the production of sperm rather than the possession of the XY chromosome) carry XX chromosomes whereas it is the females which carry XY. With this system, there is geographical variation in the strength of the genetic male determination mechanism, leading to similar variability in the ability of males to mate with females from other areas. Thus liberating large numbers of males from a foreign race with strong male determination would act like sterile male release, in this case producing large numbers of normal males but sterile intersex females. This possibility has, however, never been exploited in practice.

Competitive displacement

The replacement of one species by another is not uncommon, even without purposeful human intervention. That the North American grey squirrel (*Sciurus carolinnensis*) replaced the native red squirrel (*Sciurus leucourus*) in the UK, that a mud snail (*Ilyanassa obseleta*) was displaced by introduced periwinkles (*Littorina littorea*) in intertidal zones in New England, USA, or that the mosquito *Aedes aegypti* has been replaced in some places by *Aedes albopictus*, were all unexpected events.

However, such processes offer possibilities for pest and vector control, if the displacer is not a problem, or at least to a far smaller extent. Thus a plant or animal disease vector might be displaced by a similar insect, but one which is not a vector. Another possibility is that the displacer outcompetes the resident in an arena (e.g. for pupation sites) other than where the resident causes problems (e.g. crop or human habitation).

These are ideas that have been proposed, but no examples are as yet available, in contrast to a third approach, where at least target pests have been identified that it might be possible to displace. This approach is to try and displace the problem resident insect in one year with an insect which is also a problem, but is incapable of continued survival. Intolerance to climatic conditions is such a possibility for ensuring the later disappearance of a displacer which is itself a problem like the resident it successfully eliminated.

A form of competitive displacement has been proposed in relation to the field cricket (*Teleogryllus commodus*) in Australia. Here the genetic strain in the colder south overwinters as diapausing eggs, whereas the strain in the north breeds without such interruption. When southern females mate with northern males, only non-diapausing eggs are produced, unable to survive the southern winter. This suggests that the population in the south might be greatly reduced by the mass release of males from the north.

Apart from the numbers game inherent in mass release, there is usually no reason why the displacer should outcompete the resident. However, by selection or genetic manipulation (see later), the displacer might be imbued with the artificial selective advantage of resistance to one particular insecticide, which could then be used to drive the displacement. This would of course make this insecticide useless if it were later decided the displacer needed controlling.

Cytoplasmic incompatibility

A good example of this occurs in the mosquito *Culex pipiens*, where it has been found that some allopatric strains are often incompatible, incompatibility being either unidirectional or bi-directional. It seems to be due to the presence of a rickettsial symbiont, *Wolbachia* species, in the gonads which causes the death of the sperm in incompatible egg cytoplasm with the result that no eggs are produced.

In a field test in the late 1960s in a small village near Yangon, Myanmar, 275 000 incompatible males, comprised of cytoplasm of a strain of *Culex pipiens* from France and the genome of a Californian population, were released over 80 days. Sterility in field-collected egg rafts increased to 100% and near eradication was achieved. This success, albeit rather limited in time and space, helped generate interest in genetic control of mosquitoes, although not specifically by using cytoplasmic incompatibility.

Chemosterilization

This means the induction of sexual sterility by subjecting arthropods to chemicals, called chemosterilants, of which over 100 exist. Since they are chemicals which can be marketed, the agrochemical industry has thoroughly investigated their potential, particularly in the 1960s and 1970s.

Chemosterilants have already been mentioned in connection with the mass production of sterile insects for release; here we explore them in more detail and consider their application directly to insect populations in the field. Chemosterilants offer several advantages over using irradiation for sterilization; for example they are cheap and do not require expensive equipment.

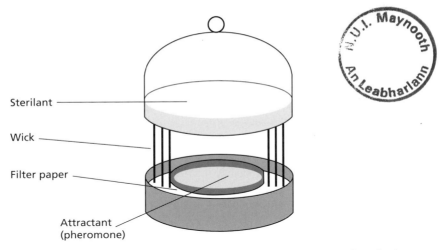

Sterilant

Wick

Filter paper

Attractant
(pheromone)

Fig. 9.3. Diagram of the type of attractive trap typically used to deploy chemosterilant chemicals in the field.

The most effective chemosterilants are either anti-metabolites such as fluoracil which act as substrates competing for enzymes in nucleic acid synthesis, or alkylating agents such as apholate, tepa and thiotepa, which replace hydrogen atoms with an alkyl group, again especially in nucleic acid synthesis. Most chemosterilants developed by industry have been alkylating agents. Sterility can be caused by a number of mechanisms, such as prevention of mating, failure to produce sperm and/or ova, killing of sperm and/or ova, and production of active live sperm and/or ova but which are sterile. This latter method is the most commonly used. Thus a typical result would be that arising from a mating of a female spider mite (*Tetranychus urticae*) with an apholate-treated male; the male offspring are viable but non-viable eggs represent the female fraction. However, there are also chemosterilants (e.g. amethopterine) where the females appear the more susceptible sex. Chemosterilants can be applied to the larval or pupal stages during rearing or introduced into adults when they are feeding. Originally it was proposed that chemosterilants could be sprayed in the field, such as on larval breeding sites (e.g. rubbish tips) of house flies or incorporated in attractant baits, but this method is today completely unacceptable because most chemosterilants are mutagenic for man. Another problem is that the most effective chemosterilants, the aziridines (e.g. thiotepa), are carcinogenic, teratogenic or even phytotoxic. If adult harmful insects are chemosterilized, then predators may also be adversely directly affected.

When chemosterilants are used directly in the field, therefore, they are localized in traps (Fig. 9.3) made attractive to the pest by a bait. As it is males that are to be sterilized, the bait is often the synthetic sex-pheromone of the female as used in the 'lure and kill' method of control with pheromones (see Chapter 10). However, plant attractants, such as mustard oil volatiles for cabbage root flies (*Delia radicum*), have also been tried.

In the laboratory resistance to apholate was induced in *Aedes aegypti* as early as 1964. Later resistance to various other chemosterilants was reported in other mosquitoes and house flies. In the 1970s the WHO, jointly with the Indian government, sponsored field trials near New Delhi, India for the control of *Culex quinquefasciatus*, an important vector of bancroftian filariasis. This was to be achieved through mass releases of adult males made cytoplasmically incompatible through exposure of pupae to thiotepa. However, the trials failed to achieve any satisfactory control, mainly due to the ability of adults to disperse over greater distances than envisaged, with the result that there was constant immigration of wild males into the target area.

Genetic manipulation of insects

The genetic engineering of natural enemies of insects to make them able to survive pesticide applications has been mentioned elsewhere (Chapter 7), as has the genetic modification of insect pathogens in order to increase their virulence (Chapter 8). In this section, we consider genetic engineering as a method of control applied directly to the harmful insects or the pathogens they transmit.

Genetic manipulation is being researched as a solution for the problem in the SIT method that it may be undesirable to release female insects. There are possibilities for transforming females produced in mass-rearing to males by continual mis-expression of sex-determining genes in females, but probably more success will come from another approach. Here genetic engineering imbues females only with an enzyme capable of converting a harmless precursor of a toxin to the toxic molecule.

Currently there is considerable interest in engineering vectors so that they are refractory to the development of parasites to man and livestock or to their transmission. Similar possibilities of genetic engineering apply to insect vectors of diseases of plants. Medically the main interest is in making anopheline vectors incapable of transmitting malaria parasites and culicine mosquitoes refractory to arboviral transmission. In pursuit of this, genetic transformation technologies are being sought that will enable the introduction of genes of any origin into pests and vectors, either temporarily or permanently, that allow the creation of genotypes that are incapable of being vectors. Because there are laboratory difficulties in inserting alien genes into mosquitoes, some believe that an alternative strategy of inserting genes for refractoriness into symbiotic bacteria or viruses is more feasible. For example, laboratory experiments have been conducted to genetically modify *Rhodococcus rhodni*, a bacterial symbiont of the triatomine bug *Rhodnius prolixus*, to express a peptide which is lethal to *Trypanosoma cruzi*, the parasite causing Chagas disease, but how feasible it will be to propagate such infections between field generations remains questionable. In the laboratory, strains of *Aedes aegypti* (vector of avian malaria) have been created that produce antibodies against malarial parasites. This was not done by engineering a transgenic mosquito but by injecting *A. aegypti* with a virus (Sindbis virus); however, how the virus will be propagated between generations has not yet been worked out. Other *Aedes* species as well as *Culex* and *Anopheles* species have been inoculated with a virus that expresses antibody genes to various pathogens. However, the World Health Organization believes it will be ten years before a vector mosquito can be rendered harmless by this technique.

There is considerable research on producing competitive transgenic strains of pests and on identifying and isolating transposable elements, such as the transposable P element found in *Drosophila melanogaster*, that are needed to drive genes that can bypass Mendelian genetics through a vector population. Unfortunately such a P element has not been found in non-drosophilids, but during the past six years the search for alternative transposable elements (transposons) in pests and vectors has proved rewarding. For example, the *Hermes* element was found in the house fly (*Musca domestica*) and has since been found worldwide in this species. This element has been used to generate stable transgenic lines in six insect species, including *Aedes aegypti,* the Mediterranean fruit fly (*Ceratitis capitata*) and the flour beetle (*Tribolium castaneum*). Other transposable elements found in, or introduced into, non-drosophilids (e.g. mosquitoes, flour beetles, Mediterranean fruit fly and pink bollworm) are *Mos1, Himar, Minos* and *piggyBac* elements.

Once genetically modified insects have been produced they must be reared and released into the target area to replace populations of vectors or, failing this, at least to increase the frequency of the refractory gene. With no driving mechanism this would involve release of enormous numbers of male insects but, with a driving force provided by transposable genetic elements or intermediate organisms such as genetically modified viruses or *Wolbachia* bacteria, many fewer engineered insects need be released. As for other forms of competitive displacement (see earlier), genes for insecticide resistance could be used to give the engineered strain an advantage, even though the same insecticide could not then be used against the strain if necessary. Another possible mechanism is 'meiotic drive'. Often one part of a chromosome is found to be passed to the next generation in a higher proportion than Mendelian genetics would predict, and inserting the desired gene into that section of the chromosome would ensure its spread.

Whether such genetically modified released insects could effectively compete with natural populations for mates remains contentious. Most researchers admit that the issue of competitiveness will be the greatest hurdle to overcome. Some advocate population suppression with insecticides prior to release of genetically engineered vectors. In addition to these scientific constraints there may also be ethical considerations, some justified but others unfounded. Genetic engineering of plants and foods has already promoted public concern, and release of 'man-made' vectors could equally be opposed as dangerous and a Frankenstein science which will lead to ecological disasters. More rationally there must be careful consideration by scientists before any vectors, genetically modified or otherwise, are released, because it is conceivable they could in some way be detrimental ecologically (see also the use of genetic manipulation in host resistance, Chapter 11). However, genetic engineering is here to stay, and there will almost certainly be benefits from genetic engineering of organisms for areas other than pest control, such as improved drugs and vaccines.

10

Pheromones

Introduction

It is becoming increasingly apparent that many behavioural activities of insects (e.g. dispersal/migration, mating, aggregation, alarm signalling and even fecundity) are under control via chemical messengers produced by individuals and liberated into the environment either as volatiles or in faeces, regurgitated food, etc. and perceived by the recipient either olfactorily or on contact. Also, chemical signals from different species, including ones from different biological Kingdoms (e.g. plants), can control insect behaviour. Those chemicals which pass messages between individuals of the same species have been given the general blanket term of 'pheromones' (see Birch and Haynes, 1982 for a good basic account) and are available to man for manipulating insect behaviour either by the use of caged insects, extracts from insects or plants or synthetic production (of the actual pheromone or a chemical 'mimic' thereof). Like all within- and between-species chemical signals, pheromones have particular advantages for pest control because they are usually highly species-specific, leave no undesirable residues in the environment and are effective in very minute quantities.

The umbrella term 'ecomone' was coined in 1977 to encompass all these communication signals, and pheromones – one type of 'ecomone' – are the subject of this chapter. The term semiochemical is also commonly used for any behaviour-modifying chemicals involved in communication between organisms, and has recently virtually replaced the term 'ecomone'. Ecomones/semiochemicals other than pheromones (allomones, kairomones and synomones) are signals between different species where respectively the transmitter benefits (e.g. predator-repellent odours), the receiver benefits (e.g. signals from prey perceived by natural enemies) or both benefit (e.g. natural enemies responding to the odour of a pest-damaged plant). These other semiochemicals may also have great potential in pest control (e.g. by manipulating the behaviour of natural enemies) and are mentioned where appropriate in other chapters, particularly under 'Other methods' (Chapter 12).

The pheromones most widely used in pest control are sex attractants, usually produced by the female of the species. Sex pheromones were first identified in moths. Naturalists have always noticed that male moths were attracted to females over long distances, and scent was always suspected. However, the quantities released are so minute that the confirmation and identification of these scents had to await the development of gas/liquid chromatography. Indeed, the word 'pheromone' was not coined until 1959. Although it was first thought that sex pheromones were peculiar to Lepidoptera, such sex-attractant volatiles are now known for insects in many different orders, a relatively late discovery being the sex pheromone of the aphid. Increasingly such pheromones are also being found in animal groups other than insects, including the nematodes and us!

The site from which the pheromone is emitted varies with type of insect. With moths, the female 'calls' from extruded glands on the abdomen, while the aphid sex pheromone is liberated from special sites on the leg.

Most sex pheromones are very specific alcoholic esters. The specificity arises not just from the actual compounds, but also from their isometric configuration, the time of release, the rate of release and the ratios of the several components often involved in the mixture forming the sex pheromone. So the same molecule can be shared by several species, one of the most remarkable examples being that the pheromone of the female elephant (*Elephas maximus*) is a molecule shared with 126 different species of moth.

There can also be differences between the pheromone 'dialect' of the same species in different locations, and isometry explains this phenomenon in the bark beetle (*Ips pini*). Beetles from California do not react to aggregation pheromone produced by their colleagues in the New York area. Both populations liberate the same chemical, ipsdienol, but the optical isomer differs. Most moths mate at dusk, but the females of different species 'call' during different time-windows as the light fades, or even at such different times as autumn or spring. Also there will be an optimum release rate for each species. The optimum release rate for codlemone (the synthetic codling moth sex pheromone) is 1.3 µg/hour. Halving or doubling this reduces the catch of males to almost zero. Many pheromones are a blend of several components.

Those components which elicit some response on their own are known as primary components; a secondary component has little effect on its own but its addition greatly enhances the attraction of the primary component, reminiscent of chemical synergists. Where a pheromone has several primary components, each may play a different role in the mating sequence, but in any case the exact balance between the components may be vital. This balance is of course a function of both quantity and vapour pressure of the various components but is usually, for practical reasons, expressed as a ratio of quantities. A good example is given by the American oak leafroller moth (*Archips semiferana*) and the ratios of the two synthesized primary components (here designated A and B for simplicity). A mixture of equal quantities catches no males. Raising the ratio of A to 60% gives maximum attraction, but a further rise to 66% reduces the catch greatly. Surprisingly (perhaps there are two genotypes in the population?) raising the proportion of A still further to 70% gives the same high catch as 60%, but with a further increase to 80% the catch falls off to nothing. The most effective component in a blend can be in minute quantities. For example, the pheromone of the pea midge (*Contarinia pisi*) has two major components which, at a 70:30 ratio, give a catch of five midges per trap per night. However, the addition of a third component at only 1% raises this catch to some 5000 midges. For some insects only one component is known and used for pest control. However, this may be just a lack of our knowledge.

As far as sex pheromones are concerned, only those produced by females have been tested in the field for pest control purposes, though increasingly we are finding that males produce sex pheromones also. However, these seem to have a much shorter effective range.

Although various types of pheromone have been identified in some medical and veterinary arthropods, including ticks, mosquitoes, triatomine bugs, bed bugs, cockroaches and phlebotomine sand flies, they have played little or no part in their control, mainly because long-distance sex pheromones have rarely been identified. Thus there are no long-range, or even medium-range pheromones emitted by tsetse flies but, as in many other Diptera, females produce a sex recognition pheromone in the cuticle which on contact induces copulation with males of the same species. These sex pheromones cannot be used to attract flies into traps because of the lack of volatility. In theory other attractants, visual and/or odours, could attract flies to targets, where a piece of string impregnated with the tsetse fly sex pheromone and also a chemosterilant would then induce settling and autosterilization. Nevertheless, there have been no worthwhile field trials on this possibility.

Fig. 10.1. Simple pheromone trap based on a plastic funnel and used for monitoring moths of *Helicoverpa armigera* in a chickpea crop in India.

Use of pheromones for monitoring pest populations

Increasingly, pheromones are being used to determine when pests enter crops, when numbers have built up sufficiently to warrant control measures being taken or to predict the correct timing for such measures. Not only can the use of pheromones greatly improve the correct timing of insecticide sprays, but the simplicity of visiting and counting insects in one or two traps in an area makes it a tool which is easy to use. The traps are usually fairly simple in design, often little more than a polythene bag attached to a funnel (usually with a vertical wall around the rim and designed for pouring paraffin, Fig. 10.1) or a small metal or cardboard roof with a sticky floor on which the insects arriving at the trap are caught (Fig. 10.2). The pheromone may be released by confining a female in the trap, but for most insects where pheromone monitoring has reached commercial practice, a synthetic pheromone is available and is released slowly from a small rubber or polythene capsule in the roof of the trap.

Unfortunately the use of female-produced pheromones means that it is the males that are caught for monitoring purposes. It is sometimes hard to relate trap catches of males to the population density of females on the crop and even harder to relate them to the number of eggs subsequently laid by

Fig. 10.2. Triangular pheromone trap with sticky floor (courtesy of Cereal Research Centre, AAFC).

females. Even for males, the relationship between population density and catch is certainly not linear, since the number of females producing natural pheromone will compete with the traps and so the proportion of males caught tends to decrease as the total population increases. 'Are there any insects?' is always easier to answer than 'How many?' Once a threshold of males caught in the trap has been reached, it may still be necessary to use temperature records and weather forecasts to predict the correct timing of the spray, which may often be directed at larvae hatching from the eggs. Of the many examples of pest monitoring with pheromones, we quote that for codling moth (*Cydia pomonella*). One pheromone trap is set up per hectare of orchard, and an average of five moths per trap triggers control with insecticides.

Use of pheromones for trapping-out pest populations ('lure and kill')

Any attractant (e.g. plant odour) can be used to lure insects into a trap where they are then killed in some way (impaction on a sticky surface, insecticide), and pheromones have been used extensively in this role.

The concept of sex pheromone traps, often treated with insecticide, to kill enough males (particularly at the beginning of the season) greatly to reduce the fertilization of the females would appear a useful control strategy. Unfortunately, mathematical models suggest that 90% of the males have to be killed before the next generation would be reduced, let alone brought down to grower-acceptable levels. Thus, for example, attempts to trap out the grape berry moth (*Endopiza viteana*) in the USA were not successful. Traps were set out on a 14-m grid, and although the percentage of infested grapes fell from 15.5 to 6.4% as a result of the traps, 6.4% infestation still exceeded what growers could tolerate. Rather more successful have been attempts to trap out pink bollworm (*Pectinophora gossypiella*) on cotton in the USA, where the economic threshold for the pest is much higher than for a pest like the grape berry moth. Traps were set out at 12 per hectare in the spring, increasing to 50 per hectare later in the season. Attack of the bolls was also monitored, and insecticide sprays were applied only when more than 10% were attacked. The technique certainly reduced the need for pesticide, which had to be applied to 45% of fields with no traps, but only to 9% of those with traps.

A massive exercise over 4000 ha of spruce forest in Norway in the late 1970s involved several thousand people to trap out bark beetles (Scolytidae). Illustrating the statement made earlier that a very high percentage of pest males in an area have to be killed for the technique to work, the Norwegian exercise did not reduce beetle populations, even though 36 million beetles were trapped over two years.

An aggregation pheromone, to which both males and females respond, is produced by the male boll weevil (*Anthonomus grandis*), an important cotton pest. Here the ability to trap out females as well as males makes the technique very successful, and it is in widespread commercial use. For example, in 1995 an area in Texas recorded 10.6 weevils per trap after the first year of using the pheromone, but by 1998 the mean had reduced to 0.04, a 99.6% reduction. Moreover, only 4.5% of the amount of insecticide used in 1995 was needed in 1998.

As well as a sex pheromone, ticks produce two other types of pheromone: (a) aggregation pheromones – which are species-specific, volatile, are produced by blood-feeding males and attract both sexes of the same species to specific feeding sites; and (b) assembly pheromones – distinguished from aggregation pheromones by being non-specific, with low volatility, they induce ticks to form clusters.

There have been several experimental attempts to use these to control ticks by the 'lure and kill' approach. Specific areas on cattle have been treated with the combination of aggregation pheromone and an insecticide such as isobenzan, toxaphene or propoxur. Beads also can be impregnated with sex pheromones and embedded in the hair coat of cattle. Males are lured to these beads and the insecticide on treated skin areas kills them as they try to mate with the beads.

A different use of tick pheromones has been to place decoys, impregnated with aggregation or sex pheromones and an insecticide, amongst pasture vegetation to attract and kill ticks off the host.

Nevertheless despite considerable interest and some encouraging results there are presently no commercial products containing pheromones for tick control, and so this approach has not been properly evaluated.

Female cockroaches produce a variety of pheromones including sex pheromones, but in many species, such as the German cockroach (*Blattella germanica*), they are non-volatile contact chemicals. However, long-range pheromones are produced by other species such as the American cockroach (*Periplaneta americana*). For this species, pheromones have been used to a limited extent in baited traps, where upon entry the cockroaches are either retained alive or killed with insecticides. When aggregation contact pheromones of the German cockroach are added, the traps become more effective. However, the main use of these contact pheromones is to mix them with insecticides such as diazinon and propoxur to make them less repellent, thus increasing the contact of cockroaches with the insecticide.

At least two sex pheromones occur in house flies (*Musca domestica*). One produced by females to attract males is called muscalure, the other produced by males remains unnamed. Muscalure, marketed as Flylure, Lurectron or Muscamone, can be combined with an insecticide to form an attractive but poisoned bait for the control of house flies. Muscalure remains an attractant for about three weeks after application. This is the only pheromone of medical or veterinary pests that is widely available commercially for control purposes.

The pheromone confusion technique

The aim of the confusion technique is to lay artificial pheromone trails, or even to saturate the crop environment with the odour of synthetic pheromone, in order to confuse the males and prevent them locating females. For this technique, pheromone traps are increasingly being replaced by smaller pheromone sources such as hollow capillary polymer fibres of only a few centimetres in length and sealed at one end (Fig. 10.3). In the USA, large 'roman ballista'-type catapults have been developed to spread the fibres over the crop from

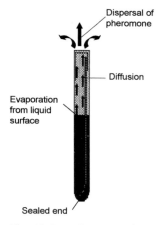

Dispersal of pheromone

Diffusion

Evaporation from liquid surface

Sealed end

Fig. 10.3.   Diagram of a pheromone fibre (enlarged).

the edge of the field, although often the fibres have to be broadcast by hand. Sometimes a brush with glue is wiped on the leaves of crop plants (Fig. 10.4) as the fibres are scattered. Another type of pheromone source for confusion is encapsulated droplets of synthetic pheromone; these can be sprayed with standard spraying equipment. Such droplets, however, have the disadvantage compared with capillary fibres that they only release pheromone for a few days.

In experiments to control plum fruit moth (*Grapholitha funebrana*) in Switzerland, hollow fibres 20 cm in length were fixed onto stakes in groups of ten. These experiments showed that, although initially the technique failed to reduce the percentage of fruit damage to an acceptable level, there was a much better result in the following year. Continuing an apparent failure into a second year revealed an important interaction between pheromone use and biological control, an interaction which is likely to have more general application. Although in the first year pesticides were much more effective than pheromone, the latter treatment allowed the natural predators to survive, and these effectively supplemented the pheromone confusion effect in the second year.

Experiments between 1976 and 1978 in California, evaluating the technique against pink bollworm on cotton, employed short 1.75-cm chopped fibres applied at 7 g/hectare every 14 days. These experiments clearly revealed one problem with the confusion technique, that control tends to break down at high populations when many natural sources of pheromone exist. None the less, the technique succeeded in reducing the number of fields which needed

Fig. 10.4. Gluing pheromone fibres on a cotton leaf (courtesy of Natural Resources Institute, University of Greenwich).

pesticide treatment and, even where pesticides were needed, there was still a valuable delay before the first application was necessary.

Similar exercises with confusion pheromone against bollworm in cotton have given equally encouraging results in other major cotton-growing regions, particularly Egypt and Pakistan. Often pheromone blends have been used to cover the range of bollworm species damaging the crop.

There is also interest in using pheromone to disrupt the mating of lepidopteran pests of stored products, with research in progress on three pyralid moths – the Mediterranean flour moth (*Ephestia kuhniella*), the almond moth (*E. cautella*) and the Indian meal moth (*Plodia interpunctella*).

It seems likely that the confusion technique can rarely be relied upon as a single control measure, but it is likely significantly to reduce the number of pesticide applications needed. At present the high cost of pheromones, inevitable with a limited demand, probably make it uneconomic to use them if pesticides also need to be used; the picture could, however, change dramatically if more farmers adopt the technique and the demand therefore increased.

Oviposition deterrent pheromones

Unusually high concentrations of female sex pheromone can inhibit or at least reduce oviposition by other females, perhaps because such concentrations in nature would signal excessive high female densities. This approach is still very much a subject of research.

Alarm pheromone

Aphids produce an alarm pheromone (β-farnesene) from their abdominal cornicles when disturbed by real or simulated natural enemy attack, and from their body fluids if the cuticle is broken at attack. Its principal effect is to increase the restlessness of other aphids. The compound is chemically simple and easily synthesized. The main application proposed has been to apply it to crops to increase the restlessness of the aphids and thus their contact with residual insecticides, which could then perhaps be applied at a lower dose, or contact with fungal pathogens of aphids. There has also been some interest in exploiting, in plant resistance breeding, the fact that some plants produce the compound, though it really requires its sudden release to initiate aphid escape behaviour.

Manipulating the behaviour of natural enemies

There is now great interest in exploring the potential of semiochemicals to manipulate the behaviour of natural enemies, and the example of indole acetaldehyde from aphid honeydew acting as a kairomone to attract natural enemies has already been given in Chapter 7. The subject will be further explored in Chapter 12, but here we limit ourselves to the manipulation of natural enemy behaviour by pheromones. Many natural enemies follow the pheromone trails produced by their prey to locate the latter; indeed, using pheromone for trapping out (see earlier) may also kill large numbers of natural enemies.

Parasitoids of cereal aphids are attracted by the aphid sex pheromone at all times, yet the cereal aphids in the crop and surrounding wild grasses are asexual and show no response. Sources of the pheromone can therefore potentially be set out in the wild grasses in the autumn to concentrate the parasitoids in aphid hosts there and not disperse wider afield. There is then the danger (see Chapter 7) that they will stay in the wild grasses when aphids colonize the adjacent crop in the spring – so then sources of the pheromone could be set out in the cereals to attract the parasitoids back into the crop.

Distribution of pheromone usage

A survey in the early 1990s, although 10 years old, at least gives some idea of pheromone usage. At the time of the survey, about 1% (some 1.3 million hectares) of the world agricultural acreage used pheromone in some form. Monitoring took place on 421 000 hectares, trapping to kill pests on 336 000 hectares and the confusion technique on the largest area, 556 000 hectares. Sex pheromones dominated, being used on 1.2 million hectares; this probably reflects the importance of their use for monitoring. Aggregation pheromone was being used on 32 000 hectares and alarm pheromone on as many as 1000 hectares. Major uses were against moths (1.2 million hectares), beetles (45 000 hectares) and flies (17 000 hectares) and cotton, fruit, forests, grapes, olives and cocoa were the only crops where pheromones were used to any large extent.

Pest resistance to pheromone techniques

The repeated use of the same pesticide has shown us that imposing a strong mortality factor on a pest population indeed leads to the initially rare genotype which survives becoming increasingly common (Chapter 5). For example, before pesticide application there may already be present at low gene frequencies a genotype with the capacity for enhanced production of pesticide-detoxifying enzyme.

It seems unlikely that there should be no genetic variation in response to pheromones within an insect population, for example in the importance of vision at close range or in the optimum pheromone blend (see the example of American leafroller given earlier). Indeed, there is already evidence of such genetic variability in pink bollworm in the USA, where a blend of two components (designated here as A and B) is used for confusion. This blend represents the peak of a normal curve of the natural distribution for pheromone emission ranging from pure A to pure B. The responses of males follow a similar distribution. The selection pressure of intensive pheromone use could therefore lead to selection for a population of males responding not to the blend, but to only one component thereof.

Thus the use of pheromone for trapping-out or confusion has every potential to select for 'resistance' to the technique in the same way as has happened with insecticides.

11

Plant and host resistance

Introduction

In relation to pest control on crops, the growing of varieties which are less vulnerable to attack than others (Fig. 11.1) or which yield well in spite of attack has many advantages. Once such varieties are available, pest control requires no extra labour and is therefore economical; moreover, the environment does not suffer from side-effects of the control measure.

Host plant resistance has been researched and used for many years for the control of nematodes and plant diseases. Even for insects, the potential of resistant varieties was already highlighted 50 years ago by Reginald Painter in the USA. Yet serious interest in host plant resistance to insects is much more recent, probably because insecticides give such cheap, effective and reliable control. The chemical control of nematodes and plant diseases, by comparison, has never been so reliable and cheap.

However, recent pressure to reduce pesticide applications has given the subject a new impetus. A further stimulus has been given by the fact that research carried out in the large internationally funded agricultural research stations (set up by the Consultative Group on International Agricultural Research, CGIAR) in the tropics to improve the major world crops has focused on plant breeding solutions for crop protection as well as for high yield potential.

'Plant resistance' poses a problem of semantics. The ability of some plants to yield better than others under pest pressure may not always be due to 'resistance' in a strict sense. As will be discussed later, so-called 'resistant varieties' may merely escape pest attack by, for example, being sown earlier, or may

Fig. 11.1.   An example of effective host plant resistance. Junction of experimental plots of the lettuce variety 'Mildura' (susceptible to the lettuce root aphid *Pemphigus bursarius*) in the foreground and the aphid-resistant variety 'Avoncrisp' behind (courtesy of the late J. A. Dunn).

be tolerant or even hypersensitive to pest attack. Some people have therefore proposed the term 'varietal control'. However, this ignores the fact that it is often possible to obtain resistance by other ways than choice of variety, e.g. by fertilizer manipulations; the term 'varietal control' is no less ambiguous than 'plant resistance'.

Plant resistance, if the result of traditional plant breeding, usually involves a quantitative enhancement of a trait already present in a plant. Southwood (1973) has concluded that plant feeding per se presents an insect with considerable difficulties; he also pointed out that plants would evolve to counteract damaging insect attack. That resistance to pests is indeed a general phenomenon can be seen in the restricted host distribution of most herbivores.

Also, small airborne insects, which have little choice where they land, apparently reject and take-off again with frequencies ranging from 50 to 95% from crop plants on which they are known as regular pests.

In marked contrast to the agricultural scene, resistance of humans or animals to arthropod attack or to the diseases they transmit has played little, if any, role in control strategies, with the exception of farmers rearing trypanosomiasis tolerant cattle (see p. 239).

Sources of variation

Most commercial insect-resistant plant varieties are the result of crop improvement by plant breeders. There are now collections (usually as seeds) of the genetic variation of many of the world's major crops. Such 'germplasm banks' for most of the world's important food crops are located at the several tropical agricultural research stations mentioned earlier. Other banks are held at national research stations and universities. Banks may hold several thousand 'accessions' (many perhaps as very small quantities of seed) derived from a variety of sources. As well as all current commercial varieties of the crop, the bank will have sought to collect locally grown peasant varieties from throughout the world. Any new intermediate hybridizations made by plant breeders during the process of developing new varieties are also added to the collections, whether or not they have displayed advantageous characters.

It is also possible to add to the genetic variation in the germplasm banks by other techniques. A very large number of new hybridizations can be produced rapidly by random outcrossing. What is needed for this technique is a male sterile line of the relevant crop. Such lines are often already available, but in any case it is practical manually to demasculate (remove the anthers) flowers of any line of crops. Such male sterile plants are set in rows between rows of any number of other lines, often gathered from a germplasm bank, each perhaps represented by only a single plant. Because all pollen resulting in seed on the male sterile plants must have come from another line, every seed on these plants is potentially a new genetic combination.

Other new genetic variability may be obtained by inducing mutation artificially. This can be done with irradiation or certain chemical treatments of the seeds. Colchicine from the autumn crocus and ethyl dimethyl sulphate are two chemicals used for this purpose. Growing plants from explants (plantlets derived from cell clusters) in tissue culture also stands a high chance of producing somaclonal (i.e. non-genetic) variation. Artificially induced variation is normally undesirable, and the seeds may even be non-viable, but there is always a chance of a useful character arising in this way.

Often the plant breeder seeks genes for improved yield under exposure to pests by using the wild parents and relatives of crop plants, rather than the variation already available in the crop. These wild relatives are known as the 'wider gene pool'. At first sight it may seem surprising that man has not selected for the resistance of these wild ancestors in breeding his crop plants by retaining the most productive types in the course of centuries. However, genetics is a science which only began with the twentieth century, and therefore most selection occurred before it was known that resistance could exist in unproductive types and that crossing could often combine resistance with productivity. Moreover, much early selection was for spot characteristics such as large fruit or large grains rather than productivity. Also, the early domestication of the wild parents involved selection for more palatable and less toxic plants, yet the chemicals that then became reduced in concentration were ones protecting the wild parent from insect attack. An example of this is the cultivated crucifers like cabbage. Levels in cruciferous weeds of the mustard oils, which give the characteristic 'cabbage flavour', may be as high as 4000 µg/g dry weight, enough to make these plants far too bitter for human consumption. To turn these weeds into crops has involved reducing these levels about eight-fold. Of course, it is pointless to make our crops unpalatable or toxic by reversing these changes, but some increase in the concentrations of such chemicals may be compatible with human consumption and can be combined with other different resistance mechanisms. Another reason for returning to wild relatives is that any resistance in the crop may have been nullified in the course of time by adapted races in the pest population or by recent importation of new pest species and races.

The further one moves from the crop species to wild relatives, the greater the likelihood that crossing barriers will prevent traditional plant breeding from exploiting the resistance mechanisms found. However, if the mechanism is a protein, then today the gene for its expression can be transferred by biotechnology. Direct methods are integrating foreign DNA into protoplasts in external electric fields or in certain chemical (e.g. polyethylene glycol) environments or by coating tiny tungsten or gold particles with the DNA and literally firing them into plant tissues (biolistics). A simpler and very widely indirect technique, which works primarily with dicotyledonous crops rather than cereals, is to replace the genes of the root pathogenic bacterium *Agrobacterium* which induce gall formation with the genes for pest resistance from another source. The now disarmed and harmless *Agrobacterium* is then incubated for one or two days with material (such as leaf segments or even isolated protoplasts) of the crop variety to transfer the gene.

The availability of such techniques has really expanded the 'wider gene pool' and genetically modified (GM) crops have made use of genes from microbial or even animal sources as well as from unrelated plants to develop pest-resistant varieties. There are more than 120 examples of gene transfer to produce plants resistant to insects, of which only 64 involve a gene of plant origin, while 15 involve genes from the Animal Kingdom. Quite a number of crops now carry the gene expressing the *Bacillus thuringiensis* (*B.t.*) toxin (see p. 228 and p. 233), as it is so effective against Lepidoptera while being harmless to most non-target organisms including man. A second toxin (lectin) has been transferred into unrelated crops from the snowdrop (*Galanthus nivalis*). Lectins give a wider spectrum of pest resistance, since they interfere with nutrient uptake in the insect; they can cross the midgut epithelium. Transforming crops with lectins is still at the research stage; high concentrations would be harmful to man. Trypsin inhibitors are other toxins being researched for future GM crops.

Location of sources for resistance

The range of variation available in the above sources has to be tested for resistance to the pest in the field through a process known as 'screening'. As often only a few seeds of each line will be available (especially if from a germplasm bank), the screening process usually has a 'negative' philosophy – each little row of unreplicated lines is compared with a known susceptible plant. Only those relatively few lines that show greatly reduced damage are tested again in further trials. In each trial, those clearly susceptible are eliminated from the trials; they may earlier just have accidentally escaped being attacked. Eventually a small number of lines, which have repeatedly yielded well in the presence of the pest, will be set out in larger areas and the level of resistance determined more accurately. As potential yield of resistant lines at this stage may be low (see 'Yield penalty' later) the resistance is best measured by a comparison of the same line with and without protection from insecticides. Clearly the smaller the difference, the more resistant the line, since only little is gained by using an insecticide.

A long process (perhaps taking 15 years) of repeated back-crossing to an agronomically adapted and high-yielding variety is then involved to perhaps produce a new pest-resistant commercial variety of the crop. The process can, of course, be somewhat shortened by transgenic techniques if the mechanism is of an appropriate type; even so, several years are still required to bring a new variety to market.

In sub-Saharan Africa resistance to trypanosomiasis, as exhibited by trypanotolerant cattle, has evolved through natural selection by constant challenge
to infection over many generations. This innate resistance can be enhanced
by exposure to trypanosomiasis transmission, especially in young cattle. Such
cattle can be kept in areas with low levels of disease transmission, but much
depends on the virulence of the different *Trypanosoma* species and even strains
within a species. The nutritional status of cattle also affects their degree of tolerance (resistance); poorly fed animals or those trekking long distances for
grazing or water are likely to suffer more from trypanosomiasis than well fed
and relatively sedentary beasts.

The classification of resistance

Painter (1951), one of the earliest workers on plant resistance to insects, divided
plant resistance phenomena into three categories, a classification which is still
extremely useful and widely used.

Antixenosis (a term which has replaced Painter's first category that he called
'non-preference', though it is not fully identical) refers to plant properties
which cause avoidance or reduced colonization by pests seeking food or oviposition sites. *Antibiosis* (Painter's second category) directly or indirectly affects
the pest in terms of survival, growth, development rate, fecundity etc., so
that population growth is reduced. Painter's third and final category, *tolerance*,
refers to a reduced plant response (usually in terms of yield loss) to a given
pest burden. It is important to realize, however, that the overall resistance of
a variety is very often a combination of two or even all three of these phenomena. Although tolerance would seem ideal in that there is no selection
pressure on the pest to adapt, it has the danger that farmers growing such
varieties would allow pests to multiply unhindered. Most workers therefore
consider antibiosis as the best type of resistance, preferably coupled with some
antixenosis.

A second more recent classification, based on genetic concepts and the
occurrence of biotypes of plant disease organisms capable of overcoming the
resistance mechanisms present in plants, is equally applicable to animal pests.
The basis of classification is Flor's (1942) 'gene-for-gene' hypothesis, which
states that genes in the plant are matched by corresponding 'virulence' genes in
the disease organism for breaking the resistance. Resistance based on a single
gene is therefore likely to be broken, as it will involve selection for just one gene.
Also a variety with a single gene for resistance will be qualitatively fully susceptible or fully resistant to different races of the disease, depending on which

races possess the appropriate virulence gene. Such resistance is termed 'race-specific' or 'vertical' resistance (the latter term from the shape of a histogram of how different races of the pest or disease perform; Fig. 11.2, top graph). By contrast, resistance based on several genes will not impose a single selection pressure and should therefore be longer-lasting. The expression of resistance to different races of a pest or disease races will be a quantitative summation of which virulence genes the race possesses, leading to 'race non-specific' or 'horizontal' resistance (Fig. 11.2, lower graph). Transgenic techniques have further moved the emphasis towards genetic concepts. However, as pointed out earlier, resistance can be induced without a genetic change. Moreover, that plant resistance usually involves new or resurrected gene combinations should not obscure the principle that the insects respond to phenomena more 'tangible' than DNA sequences. A resistance is only effective through some chemical, physical or anatomical expression of the gene combination. It is therefore possible to identify mechanisms of plant resistance and relate them to types of control (i.e. the field expression of the mechanism on the pest population).

Mechanisms of plant resistance

In Fig. 11.3, an attempt has been made to arrange the mechanisms of resistance in order of stages in the pest infestation at which they appear most effective. The aim of plant resistance is to reduce the losses in yield caused by pests. This is clearly achieved when no pest attack occurs, and the variety then succeeds in *escaping* the attack. More usually, a variety is attacked but may suffer less attack than the susceptible variety, because it is in some way truly *resistant*. Improved yield may sometimes also be gained from varieties showing a high degree of susceptibility. Some *tolerant* varieties seem able to yield well in spite of pest infestation and yet others are so susceptible as to be *hypersensitive* and collapse locally or entirely under attack; this collapse prevents the pest multiplying and spreading through the crop.

For many resistant varieties, the mechanism is unknown or poorly understood, but some mechanisms which have been identified, with selected examples, are listed below.

Colour

Insects are often affected by the wavelengths reflected from plant surfaces. The light green colour of 'Spanish White' onions seems to deter thrips from

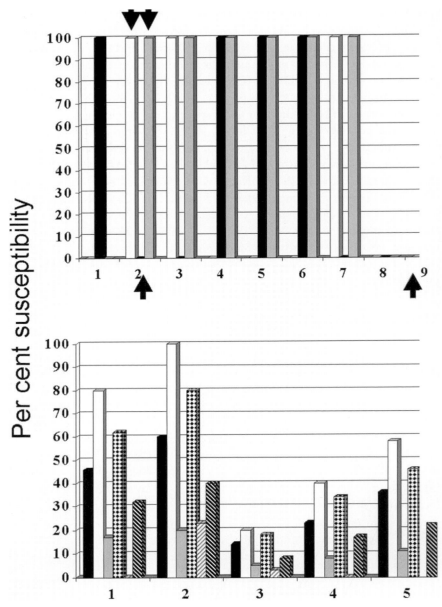

Fig. 11.2. Above, vertical host plant resistance of race-specific cultivars (1–9) to three pest biotypes (differentially filled columns). The arrows show that although the biotypes respond very differently on cultivars 2 and 9, the response is either total susceptibility or total resistance. Below, horizontal resistance of race non-specific cultivars (1–5) to six pest biotypes (differentially filled columns). Most responses are intermediate between 0 and 100% susceptibility.

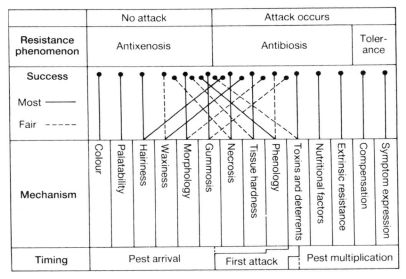

Fig. 11.3. A classification of plant resistance mechanisms based on host plant influences on insect populations (from van Emden, 1987; courtesy of Academic Press).

colonizing the plants, and the apple sawfly (*Hoplocampa testudinea*) is attracted by a high ultraviolet reflection from apple flowers and therefore oviposits more on white-flowered varieties. An interesting example is red cabbage varieties, because these represent the rare case of the insect 'getting it wrong'. The red varieties are avoided both by the cabbage aphid (*Brevicoryne brassicae*) and the small cabbage white butterfly (*Pieris rapae*) yet such varieties appear to be more favourable than normal green ones for the reproduction of aphids and survival of the caterpillars.

Palatability at host selection

Very often insects select plants which are botanically related and discriminate for characteristic secondary compounds. Some categories of these compounds are isoprenoids, alkaloids, protease inhibitors, glycosides, tannins and stilbenes. It can be shown very simply that cabbage caterpillars will not feed on leaves such as bean or lettuce, but will do so if the leaf surface is painted with the alkaloid mustard oil, the characteristic flavour compound of plants in the cabbage family. There is no evidence that such secondary substances play any role in the nutrition of the insect; indeed, the same substances deter feeding by many other insects, presumably the vast majority for any particular plant species. Breeding for high levels of these compounds may deter the host-specific insects, but probably make the plant toxic or unpalatable for man and his animals. The alternative strategy, breeding for low levels, is fraught with danger. It has been tried at least twice, in both cases the motive being, not to increase plant resistance to pests, but to make the crop residues less toxic to cattle. In both cases other pest problems were increased since a deterrent chemical was being reduced in concentration. Although low gossypol cottons were less attractive to the boll weevil, *Anthonomus grandis*, polyphagous insects such as the bollworm *Helicoverpa armigera* (normally a pest of cotton anyway) and blister beetles (Meloidae) became more abundant on the crop. Similarly low-allylisothiocyanate oil seed rape fell prey to several polyphagous pests which do not normally attack this crop.

One possible approach is to use genetic modification to introduce a novel secondary compound from an unrelated source, provided this is compatible with palatability and safety to humans.

A rather different mechanism of antixenosis is that some plants produce aphid alarm pheromone, which might be used to deter colonization by arriving aphids (but see Chapter 10).

Hairiness

An increased density of hairs on leaves may deter oviposition by small insects. In the early 1920s, the South African hairy cotton varieties were used in breeding programmes to confer resistance to leafhoppers (Cicadellidae) to other cottons, However, later work suggested that hairiness is not itself the mechanism of resistance, but is merely closely linked genetically to a chemical characteristic. Hairiness in cucumbers confers resistance to whitefly (Aleyrodidae). Hooked hairs on certain bean varieties have been shown to trap landing aphids and leafhoppers, but currently most interest centres on the glandular hairs found in wild relatives of potatoes. When small insects such as aphids break the hairs in walking over the leaf, the broken hair exudes two fluids which mix and then harden on the insect's legs and mouthparts.

Waxiness

Several workers have found differences in susceptibility to a variety of pests in glossy and waxy leaf types of cabbages and cereals. Sometimes, as in the

Fig. 11.4. A comparison of Frego bract (left) with a normal cotton variety (right) (from Maxwell and Jennings, 1980; courtesy of John Wiley).

case of flea beetles (Alticinae), it is the waxy type which is resistant; for other pests, such as aphids, whitefly and caterpillars, it is the waxy variety which is susceptible. Such resistance of non-waxy varieties to pests has been linked to higher levels of certain chemicals in the plant's cuticular wax, even though the layer of wax is thinner.

Major morphological characters

Here the best examples come from the extensive cotton breeding programmes largely with the moth *Helicoverpa* as the target. Some resistance has been obtained by breeding for narrow twisted bracts ('Frego' bract, Fig. 11.4) and the absence of leaf nectaries (nectaryless varieties), the latter originally bred to reduce sooty moulds (*Cladosporium* species) which develop where the nectaries have dripped nectar onto the leaf below. Both characteristics reduce the attractiveness of cotton to the moth, which is attracted by the low emission of gossypol from the leaf nectaries and which probably finds the Frego bracts an insecure foothold while ovipositing. Another example is the resistance of cowpea varieties with long peduncles and erect pods to the legume pod borer (*Maruca testulalis*). The resistance arises because the moth oviposits

Fig. 11.5. Plant resistance in cowpea to the legume pod borer (*Maruca testulalis*). Left, susceptible variety with pods drooping and touching; right, resistant variety with erect and separated pods (H. F. van Emden).

preferentially wherever pods are in contact with each other or the foliage (Fig. 11.5).

Gummosis

Many plants protect themselves against wounding by exuding gums, latexes and resins. In conifers, differences in resin flow have been implicated in the resistance of some pine species to attack by pine shoot moth (*Rhyaciona buoliana* – Tortricidae), and some legume varieties produce gum from the pods when damaged; this seems to drown young bruchid beetle larvae attempting to penetrate the pod wall.

Necrosis

The word means 'death'. The plant is so sensitive to attack that immediate death of the affected tissues or of the entire plant ensues. A rapid local necrosis wherever aphids probe with their stylets is a known resistance mechanism of apple against some aphid species. This kind of local necrosis can confer resistance to other forms of attack by cell penetration, and so other sucking insects, many fungi and nematodes will also be affected. Even whole plant necrosis can be a 'resistance' mechanism, attacked plants collapsing and thereby limiting the spread of infestation through the stand. Although each attacked plant is lost, this may not matter in crops such as cereals which, because of inter-plant competition, show a yield plateau over a range of sowing densities around the commercial rate (see also p. 280, but also p. 145 in relation to precision drilling). In the last 20 years, considerable interest has focused on the existence of damage-induced changes in plants which confer resistance to a subsequent pest attack. Chemically these induced changes seem similar to local necrosis, but as yet this phenomenon has not been exploited in resistance breeding. However, the principle has been adopted in some GM crops, where genetic engineering has been used to ensure that expression of the toxin which confers resistance is not 'constitutive', but only switched on when the plant is damaged.

Tissue hardness

Rapid cutinization of epidermal cells and rapid cork formation in seedlings often protect fast-maturing varieties from the many pests which concentrate their attack on young tissues. Differences in hardness of plant parts between varieties have been found important in control of cabbage root fly (*Delia radicum*), wheat stem sawfly (*Cephus cinctus*) and the rice stem borer (the moth *Chilo suppressalis*). Work on *Chilo* revealed that silica in the surface tissues blunts the mandibles of the tiny invading larvae, and silica content is also the basis of resistance of rye grass (*Lolium perenne*) varieties to the frit fly (*Oscinella frit*). However, it is not only the total silica content but also its distribution which determines the plant resistance. Some rice varieties, for example, have lots of silica grains in the tissues, but they are arranged in widely spaced rows that allow the larvae to invade the stem through the unprotected gaps between the silica rows.

Phenological resistance

This is often called 'field resistance' or 'pseudo-resistance', since the so-called 'resistant varieties' are not really resistant at all. They only appear to be resistant because they are at a less susceptible stage when pest attack occurs, either because of their rate of development or management by the farmer. Thus early flowering pea varieties escape injury from pea moth (*Cydia nigricana*), although if sown later, so that flowering coincides with the pest's presence, they are just as susceptible as other varieties. Broad bean varieties apparently resistant to bruchid beetles are not in flower at the time the females oviposit, and the number of eggs laid per female is minute in comparison with that of beetles which can obtain adult food in the form of nectar and pollen to mature a large number of eggs.

Toxins and feeding deterrents

The production by some plants of what may loosely be referred to as 'toxins' has been closely examined in the development of plant resistance to plant pathogenic fungi; phenols seem to be particularly widespread and useful fungistatic compounds in plants. They have also been implicated in the resistance of some apple varieties to root lesion nematode (*Pratylenchus penetrans*), and are probably involved in the local necrosis that occurs in response to wounding by insects (see earlier). Relatively little use has been made of the toxins produced by plants in antibiosis, presumably because the pests are mobile and show an avoidance reaction at the palatability stage. There are very many potential toxic and feeding-deterrent compounds in plants. These may be inorganic (e.g. selenium) or metabolites such as non-protein amino acids or secondary compounds (listed under 'Palatability' above). An example where secondary compounds affect feeding rather than host acceptance (see palatability) is that survival of larvae of the corn borer *Ostrinia nubilalis* as well as of the cereal aphid *Rhopalosiphum padi* is reduced in maize varieties high in the quinone DIMBOA. Grafting experiments have demonstrated the presence of leaf-synthesized toxins in some legume varieties that repel bruchids and that aphids seem unable to detect. Therefore the aphids ingest the lethal toxin. Much research on pest-resistant GM plants (including the incorporation of the *B.t.* toxin) aims to make the plants 'insecticidal' rather than repellent. This includes trypsin inhibitors, and some years before GM research started in earnest, a cowpea variety was found which, because the seeds contained a trypsin inhibitor, was resistant in store to the very serious storage pest, the bruchid *Callosobruchus*.

Nutritional factors

Southwood (1973) regarded the high carbohydrate and low nitrogen status of plants as a major hurdle for insect nutrition. In spite of a great deal of knowledge of insect nutrition and how this can be varied through the host plant, little is known of its importance in influencing the resistance of crop varieties to pests. Some information is, however, available on aphids, which mainly feed in the phloem. Phloem contents can be altered without necessarily affecting the nutritional value of the plant for man and his domestic animals. Extensive work in Canada has linked the susceptibility of pea varieties to the pea aphid (*Acyrthosiphon pisum*) with the soluble nitrogen content of the plant, particularly the amino acids. 'Perfection' gave much denser amino acid chromatograms than 'Champion' and on another resistant variety ('Onward') the aphids were comparable in growth with aphids fed on a susceptible variety but starved for 10 hours daily. Work at Reading University has linked resistance of brassicas versus two aphids, the cabbage aphid (*Brevicoryne brassicae*) and the peach potato aphid (*Myzus persicae*), to changes in the amino acid spectrum. Different amino acids affected the two species and, whilst some amino acids were favourable to aphid increase, others were unfavourable. Further work at Reading looked at a wide genetic variation in wheats, taking hexaploid varieties from the UK and Iran as well as a tetraploid primitive wheat. The published correlations of resistance to cereal aphids with either total phenolic compounds or just DIMBOA, each determined on a narrow genetic range of wheats, did not hold across the range of wheats studied at Reading, and a much better prediction came from the amino acid spectrum.

Work on a biting insect, the pea and bean weevil (*Sitona lineatus*) has shown correlation of resistance to the pest with low nitrogen : reducing sugar ratios.

Extrinsic resistance

This is the interaction of plant variety with other sources of pest mortality, the most common one being biological control. Probably the best known example is that of open-leaved crucifer varieties, which make it much easier for parasites to find their cabbage caterpillar hosts. An interesting example from Australia concerns the ladybird *Cryptolaemus montrouzieri*. This ladybird is normally phytophagous and feeds at cotton leaf nectaries. However, on the nectaryless varieties (see earlier) it becomes a predator of young bollworms (*Helicoverpa armigera*). A further very crop-specific example is that of the 'leafless' pea varieties, which have a profusion of tendrils and stipule-like expansions on the stems to replace the photosynthesis of the normal leaves which have been bred out of these varieties. The motive was to develop varieties resistant to mildew, but pea aphid problems were also greatly reduced. Although there was little difference in the population growth of the aphids on normal and 'leafless' varieties, ladybirds were able to hold onto the tendrils of the latter, whereas they tend to fall off the smooth normal leaves. This ability to hold onto the 'leafless' varieties greatly increased their consumption of aphids, since they were not spending time climbing up from the ground to regain the plants.

The very important general positive interactions between plant resistance and biological control are discussed under Integrated Pest Management (Chapter 13). However, the existence of these interactions means that, in the field, an extrinsic component is likely to be added to the intrinsic level of plant resistance for many varieties.

Compensation

Tolerance to pests often results from some form of compensation. Whole plants in a crop or different parts of the same plant can compete with one another, and removal of some of this competition will enable what survives the attack to grow larger. Thus early destruction of cereal plants by wheat bulb fly (*Delia coarctata*) is compensated for by extra growth in the plants around the victims of the fly attack.

Full compensation for leaf damage can occur when leaf area exceeds the optimum required for providing assimilates for the marketable unit (e.g. root crops). When the reverse is true, i.e. the marketable unit is produced to excess in relation to the source of assimilates (e.g. many fruit crops), natural shedding is likely to occur. Some replacement of this shedding by insect attack will therefore not necessarily affect the final yield. Many legume varieties can replace lost reproductive structures until they have produced pods and set seeds. Provided that water and nutrients remain available, there will be a harvest, albeit delayed, in spite of insect attack.

Symptom expression

One of the classic examples of tolerance to insect attack is the reaction of some tea varieties to the shot-hole borer (*Xyleborus fornicatus* – Scolytidae). Here the damaging symptom is that branches attacked by the borer break off as tea pickers push their way between the bushes. The tolerant varieties prevent this symptom by producing a wound-healing support for the affected branch.

Some cowpea varieties do not react severely to the toxic saliva of leafhoppers, which normally causes the debilitating 'hopperburn' symptom. Hopperburn is far more serious than the removal of assimilates by the hoppers' feeding. Another example of tolerance relates to some wheat varieties, which do not respond to aphid attack by twisting and curling of the flag leaf (the leaf that is most important for producing photosynthates to fill the grain), the damage that occurs in most varieties.

The problems of using plant resistance

The use of plant resistance is not without disadvantages, and the degree of these is likely to vary with the mechanism of the resistance. Thus there are several disadvantages associated with resistance based on a single strong chemical toxin of the kind used in GM pest-resistant crops. However, the resistance mechanism transferred is far more important than the technique of transfer, and the disadvantages of GM crops therefore will also apply to traditionally bred varieties if their resistance mechanism is similarly an effective toxin.

Yield penalty

Most mechanisms of plant resistance appear to involve some diversion of resources by the plant to extra structures or production of chemicals. Thus it is by no means certain that any gene for resistance can be incorporated into high-yielding varieties without some sacrifice in yield. Most entomologists working in plant breeding programmes will confirm that they find resistance in varieties with relatively low yields, and that later the high-yielding progeny of breeding programmes (that have been trialled with insecticide protection) they are asked to retest all appear highly susceptible! There is, however, little quantitative evidence that plant resistance inevitably imposes a yield penalty, even though costs associated with some chemicals in plants have been calculated and vary from 0.01% of photosynthesis (for a protease inhibitor in tomato) to as high as 24% (for a triterpene in the Alaska paper birch, *Betula resinifera*). The matter is certainly sufficiently important to warrant further research, especially since pest-resistance in GM crops is likely to be not only chemically based but also with high expression. Here it is advantageous if the GM crop can be engineered not to express the toxin until damage occurs.

Health hazard

The use of toxins as resistance mechanisms is clearly a potential problem, especially with the novel toxins and their high expression in GM plants. Some years ago, a new variety of potato with a high alkaloid content was sold and caused illness in consumers before it was withdrawn. There could also be detrimental changes in levels of vitamins and other essential nutrients. It is a little paradoxical that a toxin only needs safety testing if marketed as an insecticide! The late Kenneth Mellanby always enjoyed claiming that if rhubarb were not natural it would have to be banned because of its high oxalic acid content.

Damage to biological control

It was mentioned earlier that plant resistance is often magnified by the greater impact of natural enemies, and that this would be discussed under Integrated Pest Management (Chapter 13).

Such a beneficial interaction is not to be expected where the plant resistance is based on a toxin, which may be ingested by natural enemies while feeding on prey. Quite a number of cases of mortality or sterility of natural enemies on resistant varieties can be found in the literature, and ladybirds (Coccinellidae), hover flies (Syrphidae), lacewings (Chrysopidae) and parasitoid Hymenoptera are all involved. Several examples stem from the literature of soybean varieties, where resistance to caterpillars is due to a toxic chemical in the plants.

Apart from such chemical damage, another unfortunate consequence for parasitoids is that pests on resistant varieties (and not just where chemicals are involved) tend to be of reduced size, and this in turn results in parasitoids also of smaller size and fecundity.

The potential danger to natural enemies of GM pest-resistant crops has not yet shown itself, since the *B.t.* toxin used for the first such crops is very selective and does not harm parasitoids or many predators, though there is some toxicity to lacewing larvae. However, as well as being selective, the *B.t.* toxin is also highly effective, so that little biological control is needed against the pest, or even possible! Usually a very few surviving caterpillars can be found, since the seed cannot be guaranteed 100% pure and typically 1% of the plants may be non-transgenic. But however redundant biological control becomes in the GM crop, it remains important to know the direct effects of such toxins on natural enemies, since biological control may be needed again when the pest develops resistance to the toxin in the GM crop (see later).

The conclusion to be drawn here is that, in a pest management programme which uses multiple restraints against the pests, it is the minimum amount of plant resistance necessary to achieve the desired effect which is also the optimum. This conclusion is equally relevant to the other disadvantages of plant resistance described in this chapter (particularly any yield penalty of the plant resistance, the biotype problem and tolerance to traditional insecticides).

Problem trading

It commonly happens that resistance to organism A is linked with susceptibility to organism B. Thus one student practical class in the Horticulture Department at The University of Reading ranked apple varieties in order of resistance to red spider mite (*Panonychus ulmi*), and exactly the reciprocal order was obtained by another class independently examining the distribution of apple mildew (*Podosphaera leucotricha*) disease! Any anatomical or physiological change in a plant is likely to have a different effect on different organisms. Several examples are known, such as cotton with the narrow twisted Frego bracts. Such varieties are resistant to bollworm (*Helicoverpa armigera*) but highly susceptible to plant bugs. The wilt-resistant lucerne varieties, when introduced in the USA, proved especially susceptible to the spotted lucerne aphid (*Therioaphis trifolii*). The breeders' answer to such problems is to point to the possibility of combining ('pyramiding') resistance genes to different organisms where this appears necessary. However, this greatly complicates and slows the breeding of a new variety. Moreover, pyramiding genes may not even be an option. Since, for example, hairy cotton varieties are resistant to leafhoppers but it takes smooth ones for resistance to aphids, it is hard to see how a plant could combine both resistance characters!

An interesting example of problem trading has arisen with GM cotton. The transgene coding for production of the *B.t.* toxin is intentionally very effective against bollworm in order to eliminate or at least greatly reduce the use of insecticide against this pest. However, these same insecticides also happened to control *Lygus* bugs. As a result, high numbers of these have at times forced the re-introduction of at least some of the spraying the *B.t.* crop was intended to replace!

Biotypes

In theory, resistance of a plant variety is no more a permanent control of an insect pest than is an individual pesticide. Both control measures exert a selection pressure on the pest, and the minority strain not affected by the control may then become more common. 'Breakdown' of resistance, by the emergence of tolerant 'biotypes', is familiar to plant pathologists (e.g. cereal rusts) and nematologists, and has often appeared as rapidly as resistance to chemicals. However, the biotype problem has arisen remarkably infrequently with insect pests. There are only about 25 examples which could be cited of the biotype problem in relation to plant resistance to insects, and even so many of these examples derive from purposeful testing of other strains of pests rather than any breakdown of the resistant variety in the field. There are several reasons for the relative rarity of the biotype problem with insects as compared with pathogens. These include the frequent association of antixenosis with antibiosis, the presence of several mechanisms operating simultaneously even where one mechanism was originally proposed and a relatively slow insect generation time compared with pathogens, decreasing the frequency of selection events. It is worth noting that the biotype problem has arisen most frequently with aphids, which have very short development times and so many generations a year.

Where resistance is based on a single gene, a tolerant biotype may appear quite rapidly, and this is seen as an imminent and major danger for GM pest-resistant crops. Without waiting for signs of such tolerance, a resistance-delaying strategy of leaving areas (refuges) of non-GM varieties on the farm has already been introduced for some of these crops by demand of the US Environmental Protection Agency (EPA), so that susceptible insects are available to mate with any *B.t.* crop survivors and so dilute their tolerance. These refuges are typically 20% of the area of the GM crop if sprayed with insecticide, or smaller (only 5%) if unsprayed.

The refuges are usually separate from the *B.t.* crop, but may also take the form of a within-crop seed mixture. This form of 'refuge' might cause problems with insects which move between plants as they develop, since the amount of toxin they ingest could well be reduced to a sublethal level. However, such within-crop refuges may prove necessary where pests mate before dispersing (i.e. before reaching any refuge outside the crop). Only time will tell if the refuge strategy proves effective; although field experiments look promising, models differ in their predictions. There are also potential problems where generations of the target pest are desynchronized between *B.t.* crop and refuge because of a longer development time of the pest on the former.

An oft-quoted example of the failure of a monogenic plant resistance is that of the release of the rice variety IR26. This variety showed a high level of resistance to both the brown planthopper (*Nilaparvata lugens*) and the green rice leafhopper (*Nephotettix virescens*), and so was eagerly taken up by farmers across large areas of South-East Asia. The resistance only lasted a few years, and a different strategy has been established of planting a mosaic of rice varieties with different resistance mechanisms in the area.

By contrast, other resistances have lasted for many years, e.g. wheat resistant to hessian fly (*Mayetiola destructor*) in the USA. Also, since the late nineteenth century, European vines have been grafted onto the rootstocks of some very old communion wine vines taken to the USA from France long ago, but which proved to confer resistance against the aphid-like phylloxera (*Daktulosphaira vitifoliae*) when it ravaged the European vines. This resistance lasted a very long time, and an adapted biotype of phylloxera only appeared (in Germany) as recently as 1993.

Tolerance to traditional insecticides

Although an important contribution to Integrated Pest Management (Chapter 13) is that pests are usually more easily killed on resistant than on susceptible varieties, the reverse phenomenon can also occur. This applies especially when the resistance mechanism is chemical in nature, again sounding a warning for GM plants.

Thus caterpillars on soybean varieties which have a chemical resistance mechanism show tolerance to the insecticide parathion, and tobacco bollworm (*Heliothis virescens*) larvae similarly show tolerance to the related methyl parathion on cotton varieties with a high gossypol content. It has also been found that the increase in dose of carbaryl needed to kill caterpillars of corn earworm (*Helicoverpa zea*) on resistant than on susceptible tomatoes is almost identical to the increase needed if the chemical responsible for the resistance (2-tridecanone) is added to an artificial diet for the caterpillars. There is another very significant piece of evidence that plant resistance with a chemical mechanism can induce tolerance to traditional insecticides. This is the demonstration that raised levels of enzymes capable of detoxifying the organophosphate, carbamate and pyrethroid insecticides were found in armyworm (*Spodoptera frugiperda*) caterpillars raised on several herbs and vegetables known to contain strong secondary chemical compounds.

Variability of resistance

Plant resistance is the result of an interaction of insect behaviour and physiology with definite plant characteristics. In as much as these resistance characteristics are variable according to the age of the plant and the environment in which it is grown, a 'resistant' plant may lose its resistance and show susceptibility under certain conditions.

Plants often show a peak in susceptibility to pests at a fairly early age, with young tissues and young seedlings less vulnerable and then a steady increase in resistance as the plants age after the susceptibility peak. Aphids then show another very rapid rate of increase as soon as plants begin to flower; most gardeners will have first noticed blackfly (the aphid *Aphis fabae*) when it begins to encrust the flowering stems with its black colonies. Thus it is important to identify the appropriate stage for testing any plant material for resistance.

Many environmental factors have been reported as modifying plant resistance phenomena. Resistance is sometimes lost at either low or high temperatures, or at low light intensities. Also it is now clear that very large decreases in secondary plant compounds, resulting in loss of plant resistance, can result in glasshouses where plants are not stressed by drought or wind. Yet many researchers continue to assess plant resistance to pests in glasshouse experiments!

Major plant nutrients, pesticides and other chemicals such as plant growth regulators can also affect plant resistance. All this increases doubts about the reliability of glasshouse studies of plant resistance, and the importance of confirming resistance in a range of environments. However, the positive aspect is that there is enormous scope for obtaining resistance by techniques other than plant breeding. Obvious possibilities for inducing resistance are fertilizers and plant growth regulators. Such induced resistance circumvents all the problems and the delays of incorporating resistance with other desirable plant characteristics in a plant breeding programme.

Plant resistance and vectors of plant diseases

Aphids and other vectors of plant diseases may be more restless and show increased probing on resistant varieties. This has frequently led to suggestions that plant resistance could spread, rather than reduce, virus transmission (especially of the 'non-persistent' type that can usually also be transmitted mechanically with an infected point of a pin). However, no field example can be cited where the introduction of a variety resistant to a vector has increased the spread of a plant disease, although it has been possible to show this effect in cage experiments. By contrast, there are several examples from the field where disease resistance has clearly been reduced, even on varieties only partially resistant to the vector.

Conclusions on GM crops

The point has been made repeatedly above that GM pest-resistant crops have more serious potential disadvantages than more broadly based polygenic resistance. They are more likely to fail from the appearance of adapted pest biotypes, to damage biological control, to induce tolerance to traditional insecticides in pests and to carry a larger yield penalty. However, it must be emphasized that a contrast between GM and non-GM crops is not the right one; a better contrast would be one between resistance based on a single effective chemical and other resistance mechanisms. Even that comparison is not wholly valid, since a new technology should be compared with the technology it is expected to replace. With GM crops this is rarely going to be another resistant variety; it will usually be pesticide use. It will also rarely replace doing nothing to combat pests, yet this is the comparison often made in public debate.

The right comparison is therefore GM pest-resistance and insecticides. Then GM crops show up rather well in relation to the potential disadvantages discussed above. For example, any toxicity to natural enemies is usually small compared with the almost total mortality caused by most insecticides. GM crops also have general advantages of safety to humans and the environment not shared by many pesticides.

Vertebrate host resistance

Introduction

Antibodies, cell-mediated immune responses and cytokines play a critical role in determining the course of infection in vertebrate hosts. Consequently it is not surprising that there is interest in employing host immunity, such as by vaccination of anti-pest antigens, as a means of controlling pest attacks and disease transmission. For example, early work in the 1970s showed that when rabbits and guinea pigs were immunized with mosquito extracts, blood-feeding mosquitoes suffered increased mortality and reduced fecundity; similarly, when rabbits were immunized with extracts from stable flies (*Stomoxys calcitrans*) higher mortalities were recorded in blood-feeding flies. Initially extracts based on whole arthropods were frequently used, but more sophisticated immunological procedures have identified the arthropod tissues or their actual immunogens that induce the highest degree of immunity. Frequently, and perhaps not surprisingly, these are found to be salivary gland immunogens, or gut-derived epitopes. However, it is only with ticks than any real progress in vaccine-based control has been made. A good, although rather technical, review of modulation of host immunity by blood-sucking arthropods is presented by Schoeler and Wikel (2001). The situation in ticks is briefly described, followed by an account of use of natural tolerance of certain breeds of cattle to trypanosome parasites.

Resistance in vertebrates is selected rather than achieved by a programme of breeding, and has really only two mechanisms: antibodies and tolerance. In analogy to plant resistance, these are respectively the mechanisms of 'toxicity' and 'symptom expression'.

Resistance to ticks

Ability to control ticks without insecticides is very attractive because there would then be no insecticide residue problems in food animals. The most important tick worldwide is *Boophilus microplus;* it has been estimated that this causes annual production losses in South America and Australia alone of US$ 130 400 million. So not surprisingly control initiatives have focused on this species. There are basically two methods that might be used, firstly improving natural host resistance to ticks, and secondly vaccinating cattle against specific ticks.

Improving natural resistance
African Zebu cattle (*Bos indicus*) develop a considerably greater resistance to tick infestations of *Boophilus microplus* than do European breeds (*Bos taurus*). In resistant breeds tick infestations cause little loss in productivity. Blood-feeding female ticks on European cattle substantially reduce weight gain, whereas the same numbers of ticks on crossbred cattle (*B. taurus* × *B. indicus*) do not. In Australia the introduction and breeding of these tick-resistant Zebu cattle has reduced the frequency of insecticidal treatments. However, the protection can often be species-specific; for example host resistance to the tick *Amblyomma americana* does not confer resistance to other *Amblyomma* species.

Vaccination
Because of the increasing problems of insecticide resistance in ticks, alternative control measures are highly desirable and the development of anti-tick vaccines appears to have good potential. Although experimental vaccination against ticks was tried over 60 years ago, progress in understanding of acquired immunity has been slow. Tick infestations elicit a variety of immunological responses in their hosts, but only some may be exploited to afford protection.

Immunization of cattle with homogenates of midgut, ovaries and other internal tick tissues can sometimes provide protection against tick challenge. Initial experiments showed that ticks on immunized animals either died while still attached or they failed to complete feeding so that their reproductive potential was reduced, sometimes by 50–70%. This is because ingestion during feeding of host antibodies causes lysing of tick midgut cells, disruption of tissues and leakage of host blood to the haemocoel ('red ticks'). These results led to more sophisticated experiments. A membrane-bound glycoprotein (called a 'concealed' antigen because gut antigens used in vaccines are not secreted into the host during natural infestation) was isolated from the midgut of ticks (*Boophilus microplus*). It was then sequenced, cloned in the bacterium *Escherichia coli*, and also in cell lines of armyworms (*Spodoptera frugiperda*), and a recombinant antigen produced. When cattle were immunized with this recombinant vaccine, engorging ticks on these animals were severely damaged, resulting in a dramatic increase in 'red ticks'. Recently there has been commercial production of recombinant vaccines (TickGARD Plus in Australia and GAVAC™ in Cuba) for protecting livestock. In field trials both vaccines initially proved to be effective, sometimes reducing larval tick production of *Boophilus* species by 90% per tick generation. However, in later trials their efficacy has proved controversial, mainly because there have been no consistent results. Because a 'concealed' antigen is involved antibody titres are not boosted by natural tick infestations and booster vaccinations are required for sustained tick control. There is little knowledge about the effect of either natural or induced immunity to ticks on their ability to transmit diseases, but some preliminary experiments indicate that transmission may be reduced. There have been encouraging reports on laboratory vaccinations with other tick species, but much more work is needed before the potential of the method can be evaluated.

There might, however, be problems because some ticks mobilize granulocytes and plasmatocytes around the damaged gut areas in an attempt to repair lesions caused by the vaccine. If some ticks are successful, or partially so, in repairing such damage then resistant populations might be produced, analogous to the plant resistance problem of resistance-breaking pest 'biotypes'.

Trypanotolerant cattle

The first domesticated cattle, derived from European–Asiatic stocks, arrived in Africa in about 500 BC, and thereafter successive waves of cattle owners migrated into the continent. Either areas densely infested with tsetse flies

Fig. 11.6.   West African Zebu cow suffering from animal trypanosomiasis (nagana) showing emaciated condition (courtesy of I. McIntyre).

(*Glossina* species), the so-called fly-belts, had to be avoided otherwise cattle would become emaciated and many die, or livestock had to evolve resistance to infection with trypanosome parasites which cause animal trypanosomiasis (nagana). This is a disease that has always had a much greater economic impact in Africa than the human form, sleeping sickness. It continues to hinder rural development over much of Africa. Some 44 million cattle, as well as other domestic livestock, live in tsetse fly-infested areas and are thus under threat of nagana. About 10 million of these are trypanotolerant cattle, with pure breeds totalling 7 million and partly trypanotolerant crossbreeds 3 million.

Exotic cattle, such as European and North American breeds, succumb rapidly to trypanosomiasis and usually die. Humped breeds such as Zebu cattle (Fig. 11.6) are generally susceptible to the disease but it takes a chronic form, while humpless (taurine) breeds which comprise two main groups, the N'dama (Fig. 11.7) with their long horns and the West African shorthorns (e.g. Muturu breed) are considered to be trypanotolerant. Trypanotolerance has developed over many generations by natural selection under constant exposure to infection. Inherited resistance can be augmented by exposure to local trypanosome populations, especially when cattle are young. Even in these cattle, however, resistance is not absolute, because when they are exposed to

Fig. 11.7. N'dama cows which are trypanotolerant and in good condition (source untraceable).

a high challenge in a new area, or when they are malnourished or otherwise under stress, they succumb to trypanosomiasis. There are, however, some problems with trypanotolerant breeds. Generally they are smaller, less productive than non-trypanotolerant breeds, and nomadic pastoralists favour the susceptible Zebu cattle which are larger and can trek long distances in semi-arid conditions. In fact trypanotolerant breeds of cattle are as yet uncommon in many African countries. It may be best not to try to replace Zebu cattle with tolerant ones in the dry savanna areas of East and West Africa, but to focus on increasing trypanotolerant breeds in humid savannas and forested areas, and on commercial ranches. In fact such cattle have been introduced to humid areas in central Africa (e.g. Gabon, Cameroon, Congo) where originally there were no cattle. Trypanotolerant cattle are also useful as draught animals and their use has enabled agricultural production to increase in many areas.

12

Other control methods and related topics

Introduction

The problems of the continued widespread use of pesticides, and particularly the absorption of the words 'environmental pollution' into common vocabulary, have caused scientists to look seriously at any ideas for pest/vector control which do not involve traditional insecticides. Although the preceding chapters have discussed those control approaches which have the widest general application, there are several other methods which are, or have been, in use or have been proposed.

In this chapter we bring together this variety of unrelated methods which do not warrant a separate chapter of their own. The ingenuity of mankind in inventing ways of tackling insect problems has led to a huge variety of such methods, and we cannot pretend we can make this chapter comprehensive.

We conclude this chapter on 'Other control methods' in a different vein – attempts legally to enforce control practices ('Legislative controls'), and finally we raise some 'Other topics' which do not directly lead to the control of insects but which we believe are relevant to the topics covered in this book and should be included.

Physical methods

Such controls aim to reduce insect populations by using devices which affect them physically or alter their physical environment. Some may be hardly distinguishable from environmental or cultural controls and are frequently

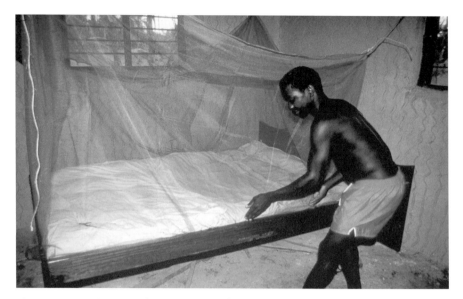

Fig. 12.1. An insecticide-impregnated mosquito net used for control of night-biting insects such as anophelines (courtesy of F. Stich and the Wellcome Trust).

labour-intensive. For example, in the early days of pest control in developing countries, handpicking and foot-crushing of larger insects (e.g. caterpillars) was economically viable and effective. The 'fly swat' is still used in many countries – no insect appears to have evolved resistance to it!

Exclusion

Mosquito nets

Nets were used in biblical times to protect people against mosquitoes, and at the beginning of the twentieth century their use was advocated by Sir Ronald Ross, discoverer that malarial parasites are transmitted by mosquitoes, as a means of protection against malaria. Cotton or synthetic fibre nets should have a mesh size of 1.2–1.5 mm to protect against mosquitoes but a smaller size of <0.4 mm if they are to exclude the diminutive phlebotomine sand flies. Nets should be firmly tucked in under mattresses (Fig. 12.1), not allowed to drape loose over beds. Torn nets, unless impregnated with insecticides (see p. 97), are virtually useless. Nets will only protect individuals against mosquitoes and other pests or vectors such as triatomine bugs, phlebotomine sand flies and bed bugs if they bite at night after people have gone to bed; they will afford no protection against day-biting pests, such as many culicine mosquitoes. A disadvantage of sleeping under nets is that they reduce ventilation and in hot climates this is unwelcome. The role of the more effective pyrethroid-impregnated nets is described in Chapter 4.

Screening

Buildings such as houses, hospitals and hotels can have their windows, doors and other openings covered with mosquito screening to exclude not only mosquitoes but other insect pests such as house flies and related flies that are often more of a nuisance in houses than mosquitoes. Screens should preferably be made of plastic and have holes 1.5 mm or less in diameter to exclude most mosquitoes. Finer mesh screens will keep out smaller insects but may considerably reduce light and ventilation. Screening should be regularly inspected for damage.

For years people in many countries have fitted anti-fly curtains over doorways of their houses to exclude flies. Such curtains are made commonly of vertical strips of threaded beads, or plastic, usually multicoloured. Such curtains may also be 'virtual'; the establishment in doorways of an air current, such as air-barriers found in the entrances of some shops, and fans mounted over doorways may help reduce the number of obnoxious insects entering.

Preventing access of insects to their breeding habitats is an obvious environmental modification method. Thus any replacement of grain storage bins etc. made of woven stems or wattle and with many holes, cracks and poorly fitting lids/doors by metal containers/silos with well-fitting covers effects an enormous improvement in the protection of any initially pest-free stored food, including against rats.

Natural covers may be encouraged – for example, the aquatic free-floating fern, *Azolla* species. These are sometimes cultivated, especially in Asia, as animal food or as a fertilizer. But *Azolla* has the additional benefit of forming a dense covering on the water surface, and so acting as a physical barrier preventing some culicine mosquitoes (mainly *Culex* species) from ovipositing or emerging. More usually synthetic materials are used. For example, pouring expanded polystyrene beads (2–3 mm diameter) down pit latrines to form a 2–3-cm thick floating physical barrier on the water surface prevents both mosquitoes emerging and more importantly from laying eggs. This provides simple non-polluting control for *Culex quinquefasciatus*, which breeds in septic tanks, latrines and other organically polluted waters (Fig. 12.2), and is a major vector of bancroftian filariasis through much of the tropics.

In the Indian subcontinent an important vector of urban malaria is *Anopheles stephensi*, a species which often breeds in water tanks sited on rooftops of

Fig. 12.2.   Pouring expanded polystyrene beads down a soak-away pit to prevent colonisation by the mosquito, *Culex quinquefasciatus* (courtesy of C. F. Curtis).

houses. Fitting them with mosquito screening would prevent breeding, but even when fitted such covers frequently become torn or are removed.

Perforated plastic covers and 'horticultural fleece' (Fig. 12.3) are available to protect young plants in the field and raise the temperature around them to produce earlier crops. The plants simply grow through the covers as they develop. These covers have proved very effective at reducing colonisation by insects attacking the crop early, and insect-borne diseases of the crop have been especially well controlled.

Before the advent of modern insecticides, the standard method of cabbage root fly (*Delia radicum*) control was to slip discs (Fig. 12.4) of tarred felt or old carpet, with a slit in one side, around the stem of cabbage plants at soil level to prevent the flies from laying eggs close to the plants. This method, like many other labour-intensive but highly effective physical methods from the past, has been resurrected in recent years by organic growers of brassicas.

Traps

Sticky traps

The use of sticky surfaces provides two further examples of techniques from the past that organic growers of today find useful. Firstly, grease bands around the

Fig. 12.3.   Horticultural fleece in position over a crop (courtesy of R. Collier and HRI).

Fig. 12.4.   Brassica plant protected from cabbage root fly by a disc of tarred roofing felt (H. F. van Emden).

trunk of apple trees trap the flightless females of winter moths (*Operophthera brumata* and some other species) when they emerge from their pupae in the soil in early winter and climb to regain the branches and twigs to rendezvous with the winged males (in case the reader wonders how the species ever left the first orchard where it appeared, it is the tiny young caterpillars which are carried long distances by the wind). Secondly, adult flea beetles (Alticinae) whose larvae feed on brassica leaves, are largely removed when caused to jump onto a plank with a sticky surface when disturbed as the plank is dragged by two operators over the surface of the plants.

Fly papers, consisting of paper or plastic tapes covered with a permanently sticky adhesive incorporating sugar, are often suspended from ceilings in houses and hotel kitchens. Flies attracted by the sugar are trapped in the adhesive. Fly papers may remain effective for several weeks before they become covered with dust and trapped flies. Although efficient, they are unsightly.

Coloured sticky rectangular traps have been used for a long time in the field, not for pest control, but for aiding spray decisions by indicating when the insects invade at the start of the season. It is only more recently, with the development of tolerance to insecticide in many pests in the glasshouse, and the widespread use of biological control in such environments, that similar traps have found a use directly for pest control. In the closed glasshouse environment, sufficient numbers of small pests (e.g. thrips [Thysanoptera] and whiteflies [Aleyrodoidea]) get caught on sticky traps that numbers of the pests on the plants are noticeably reduced. Yellow or dark blue traps are preferred by different insects, so it is common to see a mixture of cards with these colours in glasshouses (Fig. 12.5). In the USA, a new type of trap for whitefly is being tested to replace the 'yellow cards'. This trap has a green (530 nm) light emitting diode (LED), and whiteflies are caught on a sticky detachable cap. The trap has the advantage of catching very few of the parasitoids (*Encarsia formosa*) released in glasshouses against whitefly, but the disadvantage of needing nine traps to replace one card. However, each card costs US$ 1, compared with the few cents the disposable cap of the trap costs, so use of traps might well be economic over the course of time.

The whole plant becomes a sticky trap if sprayed with a sticky substance. For a short period in the late 1950s, sticky sprays (polybutenes) were applied to plants to trap small insects and mites when they emerged from the eggs, yet not to trap larger beneficial insects. Plants often reacted rather badly to these sticky films, and they are no longer used, although recently research with other adhesive sprays has apparently been initiated by industry.

Fig. 12.5. Yellow (paler) and blue (darker) sticky cards hanging in a glasshouse (H. F. van Emden).

Other traps

There are a whole variety of insect traps using an attractive principle other than the simple colour of the sticky trap; they often effect a significant reduction in insect numbers. Some such traps have already been mentioned elsewhere (e.g. traps baited with pheromones used in the 'trapping out' technique and with chemosterilants, see Chapters 10 and 9 respectively). Such information will not be repeated here. One general point about attractive traps, however, which will be made again in Chapter 13, is that most traps can be used together with insecticides to make any toxin highly selective and non-polluting.

Ultraviolet electrocutors are traps operated by a mains supply of electricity and often seen mounted on walls in shops and restaurants. They have one or two fluorescent tubes emitting ultraviolet and blue light which are supposed to attract obnoxious insects which are then killed by an electrocuting grid, their bodies falling into a collecting tray. Blow flies are sometimes attracted, but other insects, including house flies and mosquitoes, only rarely so. They are not very efficient traps.

Traps containing an attractant bait have been used for many years in cockroach control, and are still used, especially where insecticidal spraying is not feasible or when cockroaches have developed insecticide resistance. Traps can

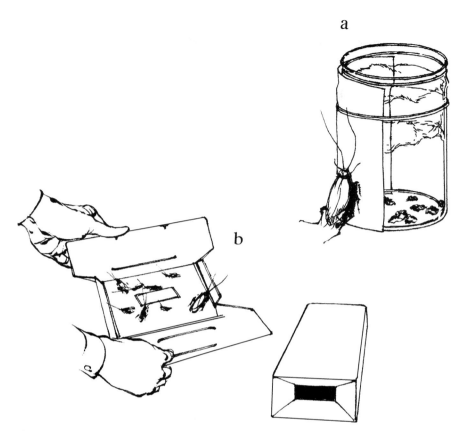

Fig. 12.6. Baited traps for cockroach control: (a) Glass jar containing food and with rough paper on the outside enabling cockroaches to climb in; the smooth glass interior prevents them from climbing out. (b) Commercial cardboard trap, baited with food or cockroach pheromone, having a sticky interior to trap cockroaches (from Rozendaal, J. A., 1997).

be home-made or bought. A very simple trap consists of a glass jar, for example a large jam-jar, with food on the bottom and with the outer surface
covered, or partially covered, with paper or some other rough surface enabling
cockroaches to climb up and enter the jar (Fig. 12.6a). The smooth glass sides
and a thin coating of Vaseline along the inside rim prevents the cockroaches
from escaping. A simple commercial trap consists of a cardboard box baited
with food or a pheromone and with the floor coated with a strong adhesive or
covered with sticky paper (Fig. 12.6b). Over 30 years ago in the USA it was
found that light-reflective panels attracted stable flies (*Stomoxys calcitrans*)

and that when these were coated with permethrin insecticide such traps attracted and killed the flies. This technique, however, has not been employed to any extent. The attractiveness of mating calls of bush-crickets (Orthoptera: Tettigonioidea) has been used in research to lure them to their doom onto the surface of a huge loudspeaker painted with insecticide. The attractiveness to slugs of beer, presented in a 'pitfall' trap, is exploited by many home gardeners and organic growers who wish to avoid broadcasting toxic slug baits.

In the above examples traps have constituted a minor role in pest control, but with tsetse flies (*Glossina* species) traps have sometimes comprised a major component of control strategies. The immature stages of tsetse flies are buried in the ground for 3–5 weeks and are well protected against insecticide spraying, which consequently has to be directed at adult resting places, such as vegetation, or to targets (see p. 98). Research in the 1970s into the response of tsetse flies to odours and colours led to the development of environmentally friendly biconical and monoconical traps visually attractive to tsetses. The lower half of a biconical trap is made of dark blue cloth and has four openings for flies to enter; the upper part is made of white mosquito netting. Trapped flies are caught in a small cage at the top of the trap (Fig. 12.7). Monoconical and pyramidal traps essentially have a cone or pyramid mounted over vertical panels of dark blue and/or black cloth. Such traps have proved effective in trapping riverine vectors of trypanosomiasis such as *Glossina palpalis* as well as the more widely dispersed savanna species, *G. morsitans*. Location of traps is important for effective control; sunny and exposed sites are usually the best as they are readily seen by the flies, the range of attraction being about 50 m. Adding acetone or 1-octen-3-ol provides a potent, cheap and easily obtained olfactory stimulant that increases trap efficiency.

Sometimes traps are impregnated with a pyrethroid insecticide which increases their efficiency because this kills tsetse flies resting on the outside of the trap without entering it.

Lethal ambient conditions

Raising the temperature to above the thermal death point for an insect has been used for many years in the routine hot water treatment of plant material to kill concealed pests such as the hover fly-related bulb flies (*Meredon* species) and eelworms (nematodes) of plant storage organs (e.g. roots, corms and bulbs). An example recommendation (for the stem nematode *Ditylenchus dipsaci*) is to soak flower bulbs for 3 hours in water at 44.4 °C between lifting and replanting. Unfortunately, generality of the technique is limited because the

Fig. 12.7.   A biconical trap used to 'trap out' tsetse populations and so control African trypanosomiasis (M. W. Service).

thermal death point of many pests is quite close to temperatures which damage the plant organ, and such narrow safety margins account for the precision of the recommended temperature in the example just given. Hot water treatment has also been used to kill body lice (*Pediculus humanus*) on clothing.

Logs left in the forest can be moved to a sunlit clearing, debarked and the piles covered with transparent plastic sheeting, so that the timber heats up from the solar radiation and 'cooks' any remaining wood-boring insects.

High temperature is also used to kill insects in stored grain when moving the grain on conveyor belts during ship or silo loading, procedures which allow the opportunity for 'fluidized bed' heating. This involves the conveyor passing through a region of high temperature long enough in relation to the speed of the belt for the heat to kill the insects. Of course, when the grain cools it has received no protection against re-infestation; also for many markets (e.g. Japan) dead insects are as much a cause for rejection of the consignment as living ones, since the corpses suggest there will be insecticide contamination in the foodstuffs.

At the other end of the temperature scale, expensive cooling systems chilling the air in insulated grain stores ('aeration') in the tropics to below 15 °C reduces or prevents pest reproduction in the stored grain.

A very effective lethal atmosphere can be created in hermetically sealed food stores by adding nitrogen to reduce oxygen concentrations to below 2%, or the more practical addition of carbon dioxide to raise the concentration of this gas to over 70%. Such techniques are known as 'controlled atmosphere storage'.

Other physical methods

A bizarre venture into physical control methods was made by a Dr Hanna of Fisons Ltd in Africa in the 1930s. He discovered that hoppers (Cicadellidae) in cotton could be suffocated very cheaply by spraying very liquid mud and blocking their spiracles (as is the mode of action of mineral oil, see Chapter 3). However, the local farmers seemed unwilling to accept that anything free could be effective. So Hanna had empty insecticide drums and labels for a mythical product called 'Hannane' shipped out to him and he filled them with the local soil. The farmers were then delighted to buy the drums very cheaply and follow the instructions "Mix with water and spray"!

Repellent chemicals are the subject of a subsequent heading, but repellence based on physical principles is discussed here.

Short wavelength light (sky) reflected from pieces of aluminium foil laid between the rows in a crop can greatly reduce the number of aphids landing on the plant by inducing descending individuals to attempt to fly upwards again. The material and labour costs of this method are high, but it has been used commercially in some high-value crops (e.g. in the cut flower industry, particularly gladioli), where it has given effective control of aphid-borne virus disease.

New synthetic cladding materials for glasshouses have been developed which block light in the ultraviolet part of the spectrum. These greatly reduce the heating-up of glasshouses in the summer. It was soon realized that they reduce the multiplication of the fungus *Botrytis*, but more recently it has been found that they also reduce the number of glasshouse whitefly (*Trialeurodes vaporariorum*) by some 80–90%. Numbers of the main parasitoid (*Encarsia formosa*) released against whitefly are also reduced, though by somewhat less (40%), so the natural enemy : pest ratio (see Chapter 13) is improved by the cladding.

Battery-operated electronic buzzers or so-called ultrasonic repellers that emit a high-pitched sound have been widely advertised in Europe, North America and Asia to repel mosquitoes. These devices are generally called 'mosquito buzzers'. Some manufacturers have claimed the sound simulates the wing beat of a dragonfly which, being a predator, 'frightens' mosquitoes away. Others have claimed that it mimics the sound of male mosquitoes and this is repellent to mated females that are host-seeking. All such claims are scurrilous; these buzzers are of no use. Some manufacturers have faced court cases in the USA and the UK and been accused of unfounded claims, and been fined.

Repellent sounds have also been researched against agricultural pests, also with little practical success. Of course, 'hum lines', rattling cans, beating drums and explosions from carbide guns or firecrackers have been a mainstay of deterring birds from feeding on crops, but against insects it has been the calls of predators (e.g. the supersonic sounds of bats to repel moths such as codling *Cydia pomonella*) that have been amplified and broadcast in orchards. Bat calls are of course inaudible to man and so are not a direct nuisance, but it has whimsically been suggested that such frequencies might set the local dogs off barking, and that they might prove more effective at reducing the abundance of wine glasses in adjacent homes than the pest population!

Behaviour-modifying chemicals (other than mentioned elsewhere)

Here we collect together a miscellany of other behaviour-modifying chemicals which do not easily fit elsewhere in this book.

Additional attractants

Pheromones are widely used attractants, and have been allocated their own chapter (Chapter 10). Other attractant chemicals have been mentioned above in connection with the physical control approach of trapping, and the attraction to natural enemies of aphids of indole acetaldehyde emanating from aphid honeydew was included in Chapter 7 (Biological control).

Also in Chapter 7, it was stated that one of the reasons why the parasitoid *Epdinocarsis lopezi* was so successful against the cassava mealybug (*Phenacoccus manihoti*) was that it was attracted to the actual location of the mealybug on the plant by the odour of cassava damaged by the pest. That the volatiles are changed when pests attack plants has been described by Marcel Dicke in The Netherlands as 'Plants crying for help'. It may well be that natural enemies have evolved to locate their prey from the changes they cause to plant volatiles rather than that the plant has evolved a different signal when damaged. Either way, such changes induced by pest attack have potential for manipulating natural enemies, as have the odours of honeydew (see above) and other faeces.

Even in the absence of pests, many natural enemies show some 'constancy' to the odour of the plant (sometimes even the exact cultivar) on which they themselves developed, giving yet further scope for manipulating their behaviour.

Finally, fruit flies (Tephritidae) are attracted by the decomposition products of fruit, and lures developed from these have been used effectively to produce attractive baits treated with insecticides and sprayed directly on the crop to improve the efficacy of the insecticide (see also in relation to the selective use of insecticides in Chapter 13).

Repellents

Application of repellents is among the most widely used methods of preventing arthropods biting people, but exactly how they prevent blood-feeding is largely unknown. Sometimes repellents are called antifeedants, but this term is usually associated more with chemicals that deter feeding of agricultural pests (see below).

When repellents are applied to the skin, their duration of protection depends on several factors, including the active ingredients, formulation, temperature and loss due to sweating or abrasion. Protection periods can range from 15 minutes to 12 hours. There are both natural and synthetic repellents. Of the former the best known is citronella oil which affords protection for only about an hour; a rather better but less well known natural repellent is the oil derived from lemon eucalyptus. Two synthetic effective repellents are dimethyl phthalate (DIMP) and diethyl-3-tolumamide (DEET), now often referred to as diethyl-3-methyl benzamide. The latter is better at repelling a wide range of blood-feeding insects, mites and ticks (and also leeches!) and is longer lasting, and has become the most widely used repellent. Repellents are often made more cosmetically attractive by formulating them as lotions, creams, foams or wax sticks, but these formulations tend to reduce their effectiveness.

The most effective formulations are usually oily liquids or aerosol preparations which normally give protection for 3–6 hours but in certain situations for 12 hours. Because of allergic reactions and skin sensitivity, application to the face (especially around the eyes) should be avoided, and repellents should not be used on infants, and also probably not on children under two years. DEET also corrodes many plastics, such as plastic watch straps and frames of spectacles.

Recently slow-release formulations of DEET employing acrylate micro-capsules in a polymer have been developed that are claimed to give longer protection. Results from trials are ambiguous, but these slow-release systems contain about half the quantity of DEET than do many other commercial preparations, which helps minimize any irritation that might be caused by DEET. There are soap formulations of DEET, usually also incorporating permethrin insecticide, and results from field trials have been encouraging, but continuous usage, especially by children, may cause allergic reactions. Furthermore it would cost US$ 4.5 a year for a person using such soap, which for poor communities might prove too costly.

There have been concerns over the safety of DEET for the last 50 years, although there is little good evidence that it is dangerous when not continually applied. But because it is increasingly used in developed countries – for example, it has been estimated that 100 million people a year in the USA use DEET, mainly against ticks and mosquitoes – some scientists argue that further tests on its safety are needed.

Recently a new repellent named Bayrepel (a piperidine derivative) has been formulated which is as effective as DEET, toxicologically harmless and does not attack plastics.

There is little evidence that use of repellents has prevented disease transmission, but during the Second World War the case rate of sandfly fever (arboviruses transmitted by phlebotomine sand flies) in American soldiers in Egypt was 2.6% in those applying DIMP, compared with 12.8% in soldiers not using this repellent. But the most convincing evidence for reduction of disease comes from trials in New Guinea where the application of dibutyl phthalate to uniforms worn by Australian soldiers was linked to a reduction in scrub typhus, a disease transmitted by *Leptotrombidium* mites.

When clothing is impregnated with DEET, protection time can be for a few weeks. However, impregnation with pyrethroids (permethrin or deltamethrin) has largely replaced DEET because protection can then extend over several months, especially if after use clothing is sealed in plastic bags. Best protection is achieved by applying DEET to the skin and wearing pyrethroid-treated clothing.

Repellents can, of course, also be used to protect animals such as pets and livestock from biting insects. For example, a topical application of a formamidine compound (Amitraz) is very effective in preventing tick attacks on dogs for several weeks. It has virtually no effect on fleas, but curiously Avon's 'Skin-So-Soft' bath oil reduces flea infestation in dogs. DEET is sometimes applied to pets to protect against ticks and mosquitoes. Occasionally repellents such as these are sprayed on livestock, but more usually pyrethroids, which are excito-repellent as well as toxic to arthropods, are sprayed on animals or incorporated into ear tags (see p. 76).

In agriculture, chemical repellents have been investigated most frequently with respect to mammals such as rabbits and bird pests, which are often difficult to control by other means. However, it has been hard to find chemicals which do more than influence choice by the animals; moreover, the compounds concerned are often very ephemeral and require frequent re-application. For a long time various 'folk remedies' have been used to repel pests with herbs and other plants, and oil of citronella (which is also used against blood-sucking flies, see above) has been claimed as particularly effective. Indeed, it is not difficult to show that leaf powders of a large number of aromatic plants repel insects in confined environments. Such powders can be easily made in the tropics from local plants and used, for example, to protect edible seeds such as cowpeas from the attack of insects in storage.

Antifeedants

The distinction between repelling an insect and inhibiting its feeding is that, in the latter case, the insect remains on the treated plant and starves to death rather than dispersing to seek food elsewhere. Some food for natural enemies therefore remains for a time. Antifeedants appear to act on the taste receptors of the insect and inhibit their perception of the stimuli to feed present in the host plant. Antifeedant properties are common in chemicals used for other crop protection purposes, for example in triphenyl acetate (fungicide), some triazenes (herbicides) and carbamates (insecticides). Carbamates may show antifeedant action at rates well below those lethal for the insect. Knowing that triphenyl acetate had antifeedant action, researchers began work on organotin compounds in general, and a number of new compounds (e.g. triphenyltin) were shown to be particularly promising. As yet, no systemic antifeedants have been discovered, so sucking insects which pierce the treated surface are not affected, and neither is there any protection of new growth or of leaf area missed by poor coverage.

Legislative controls

Legislation plays a large part in pest control and many of the activities such as the development and application of insecticides and the use of genetically modified organisms are closely regulated by legislation. Such legislation does not in itself result in any pest or vector control, and is therefore largely outside the scope of this book. However, it is important to know about it, and so such topics are given some treatment in connection with the pest control methods to which they apply. Here we describe only such legislation as forces actions which directly control noxious insects.

Legislation aimed at excluding entry

Most nations attempt to exclude undesirables (including insects!) from crossing their borders and have set up legally enforced schemes for this purpose. Economic groupings such as the European Union, as well as the lack of ecological barriers at many boundaries on continents, can make such exclusion of insects impracticable. So the intensity of exclusion efforts is closely correlated with geographical isolation of the nation and how far it is free of otherwise cosmopolitan insect problems. So Australia, New Zealand and Mauritius and to some extent California with its inland mountain chain barriers, are examples of governments that take legal steps to enforce exclusion very seriously.

Long-haul passengers on aircraft will be used to filling in declarations before they arrive at their destination that they are not carrying items such as fruit which might introduce an alien pest species into the country where they are landing, but insects of course hitch lifts on international transport in other ways, especially by merely resting amongst the structures. There is no doubt that ships have frequently transported mosquitoes across continents, such as carrying *Anopheles gambiae* from sub-Saharan Africa to Mauritius (1867–8), Brazil (1930) and Egypt (1942–4). On arrival, the mosquitoes caused malaria epidemics in these countries, while outbreaks of dengue in Guam followed introductions on boats in 1944–5 of the mosquito vector, *Aedes albopictus*.

Aircraft with their warm cabins and much shorter journey times are even more efficient than ships at spreading mosquitoes around the globe. Malaria cases have been identified in the vicinity of airports in countries or regions where there is normally no transmission. Such so-called 'airport malaria' has been reported from several countries including England, Switzerland, France, Italy, Spain, The Netherlands and Belgium, and arises due to air transportation of malaria-infected mosquitoes which bite local people. The dangers of the dispersal of medical and agricultural pests by aircraft was realized many years ago by Australia, which in 1929 became the first country to apply regulations to aircraft. Eventually a code for vector control in health was developed in the 1950s by the WHO, comprising disinfestation of aircraft after take-off and prior to landing, combined with surveillance and control around airports for at least 2 km. It is thus a common experience to have the nose and eyes irritated by the solvent of an insecticide from aerosol cans held aloft by the cabin crew passing down the aisles; of course such small drops only go upwards (Chapter 4) and we suggest that the spraying is therefore likely to be more effective at annoying the passengers than at disinfesting the plane.

Legislation aimed at containing entry

Quarantine

Quarantine is derived from the Italian word (quaranta) for forty. It was practised in Venice in 1374 in efforts to prevent plague entering the city. Those disembarking from ships were isolated in designated areas for 40 days before being allowed to enter Venice.

The objective of quarantine legislation is to prevent exotic pests and diseases becoming established in countries free of them. It is well known that the UK has strict quarantine laws whereby dogs and cats are held in 'kennels' for six months to prevent entry of rabies.

Most countries operate quarantine laws to allow inspection at the point of entry of all produce which might harbour foreign pests; these laws also enforce strict isolation of any species imported for study (e.g. for biological control research). Commercial plant material, including seeds, entering countries may require an accompanying certificate from a competent authority in the country of origin that it is free from pests; alternatively it may be quarantined by government inspectors for a thorough check before release to the trade. It may also be possible for the material to be received by a trader under a 'Post-Entry Quarantine' certificate, which specifies the conditions under which the material may be kept until it can be certified as 'clean'. Special licences with strict conditions also allow the import of exotic insects for research purposes.

Unfortunately, quarantine normally only postpones the entry of pests. The strict and largely successful quarantine as practised, for example, by Australia, causes crop varieties to be bred in the absence of some of the most important world pest problems. When such pests eventually arrive, they can cause devastating damage, as the plant breeder will not have checked his high-yielding varieties for susceptibility to these organisms. Here a classic example was the devastation of local Australian lucerne varieties by the blue-green alfalfa aphid (*Acyrthosiphon kondoi*) following its accidental importation from the USA.

Legislation aimed at preventing spread

'Internal quarantine' aims to prohibit the spread of infections out of the affected area, and is thus the prevention of 'exit' as opposed to the prevention of 'entry' (quarantine).

The International Animal Health Code periodically issued by the Office International des Epizooties (OIE) imposes restrictions on the movement of animals if they are infected with various diseases, such as Rift Valley fever (see p. 260).

Legislation has also attempted to prevent the sale of certain plants, seeds, tubers, etc. subject to named pests or diseases unless they can be certified as free of these problems. For example, 'The Sale of Diseased Plants Order' (a series of laws passed in Britain between 1927 and 1952) prohibited the sale of plants with infestations of several pests. These pests included glasshouse whitefly (Aleyrodidae), but 'The Sale of Diseased Plants Order' has probably been more honoured in the breach than in the observance, particularly at charity and car boot sales.

Legislation aimed at achieving eradication

Particularly serious problems may be subject to a 'Notification Order', whereby a suspicion that the crop pest or vector-borne disease may have appeared must be reported to the appropriate authorities, who will then impose bans on the movement of plant or animal material and/or undertake eradication. For many years, the Colorado beetle (*Leptinotarsa decemlineata*) has been a notifiable pest in Britain. The exotic noctuids *Spodoptera littoralis* and *S. litura* turned up in UK glasshouses (particularly carnation crops) in the 1970s, but were eliminated by intense pesticide treatment. They are still regularly intercepted at the docks on imported plant material, and remain on the list of notifiable pests. Another example is the silverleaf whitefly (*Bemisia tabaci*), first introduced into the UK in 1987. More recently still (in 1997) a thrips (*Thrips palmi*), which has the potential to transmit a number of wilt viruses to vegetables, was spotted on imported orchids by inspectors, and such interceptions on orchids have become increasingly common. *Thrips palmi* is now a notifiable pest.

A number of arthropod-borne diseases transmitted to humans have to be officially reported to country ministries (e.g. Ministries of Health or

Agriculture). Notifiable diseases vary from country to country, but some, such as infection with yellow fever, plague, malaria and louse-borne typhus must be reported to the WHO or the Pan American Health Organization (PAHO). Certain veterinary arthropod-borne diseases have to be reported to the OIE (see above). The OIE has two lists of notifiable problems. List A comprises diseases such as Rift Valley fever, African swine fever and many other arboviral infections that can spread rapidly irrespective of national boundaries and are of major importance in international livestock trade, and where there are strict quarantine regulations prohibiting the export of infected animals. List B includes pests such as screwworm flies, and diseases like Nairobi sheep disease, Japanese encephalitis and other arboviral diseases, that although important within countries are less so in the international trade of animals.

Legislation may also impose penalties for not dealing with certain problems, especially in relation to public health. For instance, in most countries the presence of cockroaches, flies and other arthropods as well as vertebrate pests (e.g. mice and rats) in restaurants and hotel kitchens and other places where food is prepared for public consumption can lead to fines. Local government authorities can often insist that houses are disinfested against vermin. Local authorities may have the power to inspect school children for head lice, scabies mites and other infestations, and serve notice on parents to get rid of such vermin.

Legislation to enforce prophylactic control measures

It was mentioned in Chapter 6 that many countries legally impose a 'close' season, during which no cotton may be grown, in order to break the life cycle of some major pests of the crop, though we pointed out that this was not frequently policed or enforced. This legal requirement to take prophylactic action against insects is also exemplified by 'rotation orders', as enforced in various countries at various times (e.g. sugar beet rotation to control beet eelworm in Britain).

There is other legislation requiring prophylactic action against arthropod pests that are detrimental to public health. Many tropical countries, for instance, legislate to prevent people from having water-storage pots or water tanks that are breeding mosquitoes, and there are laws to stop building contractors allowing water to collect that might provide potential mosquito breeding places. Reservoir tanks and other water collections are supposed to be fitted with covers preventing mosquitoes from ovipositing in them, but in practice very few countries enforce these regulations. An exception has been Singapore.

In the 1980s a law was passed enabling government employees to enforce community-based eradication of *Aedes aegypti* breeding places. People could be fined, imprisoned or both if there was breeding on their premises. Arguably Singapore and Cuba (see p. 268) have the most efficient control programmes in the world aimed at *Aedes aegypti*.

Other topics

This book is about the various approaches to reducing the numbers of noxious insects, and the use of a combination of methods in integrated pest management programmes (see Chapter 13). However, there are some other topics which, although not part of insect control methods, seem so relevant to us in relation to the wider picture of controlling crop pests and the diseases of man and livestock, that we include them at the end of this chapter as supplementary information. These topics are:

- A brief account of *the use of drugs against vertebrate parasites*. Control of vectors is only part of efforts to safeguard public health, and drugs are an equally or sometimes even more essential weapon in coping with diseases such as malaria.
- *The role of international organizations*. This is much more general than insect control, but international organizations – partly by their 'muscle' as providers of funds – have a powerful influence on attitudes to insect control measures and the acceptance of integrated pest management (Chapter 13).
- *Safety regulations in relation to insect control practices*. Such regulations do not kill insects or require them to be killed, but seek to prevent and prohibit unsafe materials and practices.
- *Community participation*. We can vouch from personal experience that (particularly in developing countries) effective involvement of the community, and seeking its perceptions and opinions, is as important as anything else that can be done for the improvement of insect control.

Role of drugs against vertebrate parasites

With arthropod-transmitted diseases to humans and animals the objective is disease control, not primarily vector control, although vector control may play a minor or major role in control strategies. For instance, control of sleeping sickness in sub-Saharan Africa is mainly based on drugs, with vector control playing an increasingly minor role, in part due to costs and logistical reasons.

In contrast control of many arbovirus diseases, such as dengue and even yellow fever for which there is an efficient long-lasting vaccine, relies heavily on vector control. The mainstay of malaria control in the 1960s to about the mid-1980s was insecticidal spraying of houses with residual insecticides. Now, however, there is a more integrated approach, involving prophylactic and therapeutic drugs combined with vector control measures, especially insecticide-impregnated bed nets. However, just as vectors become resistant to insecticides, so parasites become resistant to drugs. For example, malarial parasites have become resistant to a wide range of commonly used synthetic drugs, even to relatively new drugs such as mefloquine (Lariam) in certain regions. Artemisins, derived from the Chinese plant *Artemesia annua*, are increasingly used to treat drug-resistant parasites as is the combination of two drugs having different modes of action. It is hoped such combinations will delay/prevent the development of drug resistance, but drug resistance remains a problem, if not now then likely to become so in the future. Once resistance to drugs appears, the importance of returning to control the vectors increases.

Role of international organizations

We must begin with a caveat. All organizations seem to love re-organizing, finding a new acronym and then devising a wording to fit. Many of the organizations mentioned below, and later in relation to pesticide regulation, change their image rather frequently and we cannot guarantee that the acronyms we use here are current, especially as it takes several months for a book to go from manuscript to publication.

Various international and national organizations have expertise in pest and vector control, such as the World Health Organization (WHO) and the Food and Agricultural Organization of the United Nations (FAO). In the USA there are the Centers for Disease Control and Prevention (CDC), and for the Americas the Pan American Health Organization (PAHO). These organizations produce publications, such as the WHO's *The Weekly Epidemiological Record* and CDC's *Morbidity and Mortality Weekly Report* which give updates on the status of vector-borne (and other) diseases, notification of epidemics and, together with the FAO *Yearbooks of Agriculture*, valuable statistics on case numbers. These organizations also actively carry out and/or sponsor research on pests of agriculture (including insects attacking livestock) and vector-borne diseases and participate in, or advise on, control strategies. The involvement of such organizations in the regulation of pesticide usage will be discussed later.

The FAO has played a major part in promoting integrated pest management (IPM) as part of its general move towards more sustainable agriculture in developing countries by providing aid for IPM projects and mounting relevant conferences and training. The FAO and the World Bank have often achieved such developments in collaboration with national bodies such as the Overseas Development Administration (ODA) in the UK and the United States Agency for International Development (USAID).

A major player in agricultural development has been the Consultative Group on International Agricultural Research (CGIAR), albeit with the financial input of outside donors, including the international and national aid agencies already mentioned. CGIAR (as mentioned in Chapter 11) has founded a chain of large research stations in the tropics, between them carrying mandates for improvement of the majority of the major world food crops. After many years of over-concentration on plant breeding as the way forward in crop protection, they have now included IPM in their remit, and have led worthwhile progress in IPM in Africa, Asia and Latin America.

Two other important international organizations concerned with crop pests are CABI Biosciences, an international centre for biological control research though now increasingly also moving more widely to IPM, and the International Centre of Insect Physiology and Ecology in Kenya (ICIPE). The latter is more laboratory based, and has a strong interest in medical/veterinary as well as in agricultural entomology.

Other organizations that are sometimes involved in providing help with vector and disease control operations include the International Commission of the Red Cross (ICRC), Oxfam, Médicine sans Frontière (MSF), United Nations Environmental Programme (UNEP), United Nations High Commission for Refugees (UNHCR) and the United Nations International Children's Emergency Fund (UNICEF).

A significant international event in relation to insect control was the United Nations Conference on Environment and Development held in Rio de Janeiro in June 1992. Governments of developed nations were so involved that the conference has become colloquially known as the 'Rio Summit'. Many agenda items are relevant to insect control but, for the sake of brevity, we will highlight Chapter 14 of Agenda 21, where IPM is one of 12 programmes. There were several milestones to be achieved: entire policy review by 1995, mechanisms to regulate the distribution and use of pesticides by 2000, and establishing networks to develop IPM 'no later than' 1998. Sadly, but perhaps not unexpectedly, we can report that by 2003 little of all this has materialized. Still Agenda 21 does firmly establish IPM as the future for sustainable crop production.

Safety regulations in relation to insect control practices

Registration and permitted uses of pesticides

In Chapter 3, we introduced the concept of registration procedures for new pesticides and in this and other chapters have referred to the banning from further use of compounds previously permitted. Most countries have a list of permitted pesticides usually also defining the specific purposes for which they can be used, and these uses can vary from country to country. In the UK the Pesticides Safety Directorate (PSD) oversees pesticide regulations while in the USA the Federal Insecticide, Fungicide and Rodenticide Act (FIFRA) governs the use of registered pesticides, while the Act is enforced by the Environmental Protection Agency (EPA), which also takes the decisions which concern registration and use. New imported pesticides require similar governmental approval before they may be used. The heavy costs (Chapter 3) that are involved for registration for each 'use' can lead to scientifically bizarre situations. An especially good example of this in the UK has recently occurred with an excellent new herbicide for brassica crops, plants which tend to be rather herbicide-intolerant. The herbicide cycloxydim has been registered for the arable oil-seed rape crop, which is a large market for the herbicide but where a high level of weed control is not essential, but is denied to growers of Brussels sprouts, who need excellent weed control, but offer a relatively small market. A clear distinction therefore needs to be made between 'non-registration' and 'banning'. Countries may differ in what can be used. For example sale to the public of dichlorvos impregnated plastic strips (e.g. Vapona), as commonly used for house fly control, is presently no longer allowed in the UK and USA and may soon be banned in some other countries. Moreover, in some countries carbaryl and HCH are no longer recommended for control of head lice (*Pediculus capitis*) because there is some evidence that both insecticides may pose a health risk, but their use is allowed in many other countries. International organizations like the Organization for Economic Co-operation and Development (OECD), the FAO and the WHO provide guidelines on pesticide regulations. The FAO has published an International Code of Conduct on the use and distribution of pesticides, while the WHO publishes a useful list that classifies pesticides at various hazard levels according to their oral and dermal toxicities.

Application of pesticides

At the commercial level, many countries require those advising farmers on the choice and application of pesticides, as well as those carrying out the actual application, to have received appropriate training. In the UK, the requirement is to have passed the appropriate test(s) under the BASIS certification scheme. Here, for example, there are different tests for using hand-held or tractor-mounted equipment. In the UK, legislation on use of pesticides is primarily found in two sources, the *Control of Substances Hazardous to Health Regulations* (COSHH) and the *Food and Environmental Protection Act* (FEPA). In many countries, it is illegal to deviate from the 'label' (the conditions of use which may include not only dosage but also spray volume, droplet spectrum and requirements for protective clothing, provided by the manufacturer and agreed by the registration authority). In Australia, where such 'label' permissions can differ between states, we have come across farmers whose land crosses state boundaries whose use of pesticide on one part of their farm would be illegal on other parts! Note that such legislation prohibits deviation from the 'label', not exceeding label-specified doses. This is unfortunate. Although the agrochemical industry would argue that the stated doses are needed as an insurance that the effect of the application is satisfactory and that a high kill delays the development of resistance in the pest (but see Chapter 5), the 'label' does prevent application of reduced dose rates, even where entomologists, consultants and farmers would all agree that reduced rates would be desirable on pest management and environmental grounds. Equally, the legislation can prevent the introduction of improved application techniques which give better capture of the pesticide on the target (Chapter 4) and therefore the 'label' dose could be phytotoxic.

Working closely with the FAO and the WHO, the Codex Committee on Pesticide Residues (CCPR) is responsible for establishing the maximum residue levels (MRL, see Chapter 3) permitted on food. Although all industrialized countries, and most developing countries, have legislation regulating pesticide usage, in many developing countries there is ineffective implementation, often aggravated by financial restraints. Consequently some food crops destined for export have unacceptably high pesticide residues.

The issue of spray drift beyond the crop boundary has led to relevant legislation in many countries, where a variety of practices to limit such drift may be enforced. Such practices include 'no-spray' buffer zones, at least on the downwind side of the crop, particularly to protect water courses or conservation areas. The width of buffer zones may vary considerably, not only with type of pesticide but also with the type of crop and other local considerations (e.g. presence of windbreaks to filter out drifting droplets), perhaps from as little as 1 m to as much as 20 m.

Genetically modified organisms

The general public is becoming increasingly wary of genetically modified organisms (GMOs). This issue is highly relevant to pest control, since the first GM crops have been introduced for pest control motives and involve expression of the *Bacillus thuringiensis* toxin, very effective against caterpillars (see Chapter 11). Although a great deal of concern relates to human consumption of genetically modified foods, much of this concern results from misunderstandings. In contrast there is more realistic cause for concern over ecological damage to crops and wildlife that might arise from growing genetically modified crops, including the 'escape' of genes to the same species (e.g. volunteer crop plants or conspecific wild plants) outside the treated area. This is not because of the GM technology per se, but because of some of the novel characters the technique can introduce into organisms and which would be impossible with traditional breeding. The three important questions which need to be asked and answered with the ecological dimension of GM organisms, and incidentally equally with any other potential hazard in life, are (1) will there be any exposure to the hazard? (2) if so, what effect might it have? and (3) given that effect, would it matter? As with pesticide usage, countries have their own regulatory agencies such as the EPA, United States Department of Agriculture (USDA) and FIFRA in the USA, while in Europe the European Union governs the commercialization and release of GMOs on behalf of the concerns of member countries. International agencies, such as the OECD, WHO and FAO can also play important regulatory roles in relation to GMOs.

In the UK, the Advisory Committee on Releases to the Environment (ACRE), including its various subcommittees, is responsible for advising government and approving (or not) the growing of GM crops. This Committee has recently considered the information in relation to biodiversity issues that should be required of industry in submitting applications for the planting of GM crops. The main proposal is that industry should be required to make a risk assessment comparing the effects of the release (on a portfolio of biodiversity issues) with those of the whole range of alternative current agricultural practices – e.g. from high-intensity pesticide use to organic farming in the case of a pest-resistant GM crop.

Community participation

In many developing countries, whereas a farmer may appreciate that the arrival of pests on his crop threatens an economic loss, he is much less likely to realize

Fig. 12.8. Community participation. Local farmers discussing new pest-resistant cowpea varieties at a trial in Nigeria, enabling plant breeders to learn what other characters of plants are preferred locally (H. F. van Emden).

that mosquito larvae breeding in his backyard are associated with malaria, yellow fever, dengue or filariasis. Firstly, he is unlikely to know that mosquito larvae give rise to adults which bite people, and secondly even if he knows this he probably does not understand that they transmit diseases. This is because he cannot actually see the insects causing ill health in his children, but he can see caterpillars eating his crops. Even if the threat that mosquitoes pose is explained he may feel there is little he can do, or afford to do, to lessen the risk. The perception of the problem is therefore one of the difficulties encountered with community participation.

Community participation, or 'community involvement' as the WHO prefers to call it, because the word 'involvement' implies an active rather than a passive engagement in health activities, has had a short but chequered history. Politically community participation is desirable for various socio-economic reasons, but sometimes it is not the most effective procedure. For instance, a UNICEF insecticide-impregnated mosquito-net campaign in Namibia employed local village women to make the nets over many months, despite knowing that imported nets were cheaper and were immediately available.

It is essential that the cultural beliefs and the social structure of communities are taken into consideration, and these can vary enormously from area to area.

Equally, the benefits of community participation must not only be explained but also understood. Even today the poor in most tropical countries have little understanding of how the common diseases from which they suffer are transmitted.

In Puerto Rico in the 1980s, after comprehensive educational propaganda, more than 70% of the people considered dengue an important disease, and the percentage of those believing that dengue could be prevented had increased from 4.5 to 73%. Nevertheless, it proved impossible to get the community to change their habits and clean up their back yards to prevent the vector, *Aedes aegypti*, from breeding. There have been many other failures to get full and sustained cooperation of the people. Another problem is that if community actions have reduced disease transmission, people no longer fall ill, and it is perceived that the problem has gone away. The community then sees little use in continuing its efforts. Some believe that often the answer is in giving incentives, financial, economic or otherwise, if self-help activities are to be sustained.

There have, however, been some successes without giving financial rewards. Examples are getting villagers in rural West Africa to use simple traps to reduce tsetse fly populations, and getting people in Latin America to replaster the walls of their houses to prevent triatomine bugs hiding in them. In Cuba, when Fidel Castro declared that *Aedes aegypti* was 'Public Enemy Number One' it generated an *esprit de corps* in the entire community which then cleaned up the environment and removed discarded bottles and tin cans and other mosquito breeding places. Similarly, in Singapore, people helped to eliminate *Aedes aegypti* breeding places around their houses, but here success was likely because householders could otherwise be fined or imprisoned.

Despite disappointments and many failures, most national and international health agencies believe that the only long-term solution to controlling dengue in Asia and the Americas is through community participation. This approach, although logically sound, may prove very difficult for effective and sustained control.

People should not be asked to participate in any vector control programme unless there is a realistic chance that it will result in a worthwhile reduction in vectors. We must avoid the trap that just because methods are simple and attractive they will work, and ask ourselves whether community programmes for source reduction and/or environmental modifications will give as good results as more traditional insecticidal measures. With diseases like Chagas the answer could be 'yes' because relatively simple procedures like householders plastering the walls of their houses can substantially reduce biting by the triatomine vectors.

The African Programme for Onchocerciasis Control (APOC) is establishing a 12-year community-based ivermectin treatment regimen, backed up by focal larviciding, to eliminate river blindness. It will be interesting to see whether there can be such long-term sustained community involvement.

As pointed out above, the association of human and animal diseases with insects as their vectors is not as obvious as caterpillars chewing a crop plant. However, this does not mean that community participation is not equally relevant in agricultural entomology. It is essential to involve local communities in every way possible, especially to understand their priorities and motivations (Fig. 12.8). History is full of misguided imposed outside development schemes in agriculture – the story of cotton in the Cañete Valley in Peru, described in Chapter 13, is just one example. Entomologists have had to run for their lives because they mistakenly believed they were doing the villagers in Kenya a favour by controlling the giant looper caterpillar on coffee; they hadn't bothered to find out that the huge succulent caterpillars were valued as a major source of protein. We have experienced in Nigeria that there is no point in persuading a farmer to grow a higher-yielding pest-resistant cowpea variety. This is because until the roads and transport are provided to take surplus crop for sale in the nearest town market, all the farmer will do with his higher-yielding plants is to sow a smaller area!

One example of successful community-based pest management is the ICIPE Oyugis–Kendu Bay project. Here pest-resistant cultivars of several crops were combined with the training of farmers in better agronomic practice. The input of simple equipment such as ox-ploughs was by request of the farmers, and it was they who selected the varieties that fitted their production systems best, and advised the scientists on the priorities for on-site research. An overall 40% increase in crop yield was achieved.

Another example we can give is the IPM training programme in Indonesia that followed the decision (see Chapter 13) to decrease dependence on insecticides in rice. Those responsible realized that even a Presidential decree needed the additional effort of changing the behaviour of people by showing the advantages to them of such changes. The whole approach was to let individuals make discoveries for themselves and look to farmers to input their knowledge and expertise into the training efforts, especially in producing training materials. Several thousand 10–12 week training sessions were organized around 1000 m^2 'learning fields' of rice run on IPM principles, but also giving the added opportunity to teach about diseases and water management. Many thousands of farmers were trained, and this led to a huge mobilization of labour (80 000 school children with 350 000 farmers) to arrest an outbreak of white stem borer (the moth *Scirpophaga innotata*) in 1990–1991 by collecting egg masses, setting pheromone traps and releasing parasitoids. The exercise reduced damage by 93%.

These examples, whether from medical or agricultural entomology, make it clear why community participation is the way to introduce new habits and technologies. Two guiding principles emerge:

- Although the input of scientists is a prerequisite, they will be more effective if they 'listen' and not just 'preach'.
- People follow the example of their peers readily and without suspicion, especially when they see beneficial results.

The examples above all stem from the poorer developing countries. Is there possibly a lesson here for developed countries?

13

Pest and vector management

Introduction

As was mentioned in Chapter 5, man has already found situations where the insecticide 'road' has run out. Many of the available insecticides failed in the middle and late 1950s because of tolerant pest strains on cotton in Peru, on lucerne in California and on chrysanthemums in glasshouses in Britain, and already some medical/veterinary pests were becoming resistant. In her controversial book *Silent Spring*, Rachel Carson (1962) had advocated that man must choose between chemical and biological control; mankind was 'standing at a fork' of the ways. The first sentence of her final chapter ('The Other Road') in fact begins with the sentence 'We stand now where two roads diverge'. In the light of what was already happening at that time with lucerne in California, and what has happened since elsewhere, the sentence stands out as perhaps the most interestingly misleading statement ever made about pest control.

It is a useful exercise to look at the above three famous examples of failure of insecticidal control against crop pests in the 1950s and to consider the principles involved in the solutions in the light of that sentence of Rachel Carson's, 'We stand now where two roads diverge'.

It is worth noting that, at about the time of Carson's book, the authority for the registration of pesticides in the USA was taken from the Department of Agriculture and placed in a newly formed Environmental Protection Agency (EPA), and one of the first actions was to suspend usage of DDT in the USA. The EPA is a very powerful body which has restricted the use of, or suspended the use of, quite a few pesticides. The justification for these actions has sometimes been queried because at times they have made it more difficult to control vectors and prevent disease outbreaks. Similarly banning DDT and other compounds, or requirement for expensive re-registration, has made it almost impossible to control some pests. Such actions by registration authorities have resulted in farmers taking pest management as an alternative to routine prophylactic pesticide treatments very much more seriously in recent years.

The classic examples of insecticide failure on crops in the 1950s

Cotton pests in the Cañete Valley of Peru

The Cañete Valley was a vast irrigated cotton area in an otherwise dry and lifeless region. Ecologically it was an island of monoculture and therefore predisposed to ecological disaster. Once the area was regularly blanketed with insecticides, there were rapid and devastating problems. New pest problems appeared and, together with pesticide resistance, caused a yield crisis as early as 1955. In 1956 a legislative package was introduced. One element was compulsory crop rotation or a mixture of crops on every farm (Fig. 13.1) so that, whenever pesticides had to be used, there would be unsprayed refuges for natural enemies. Secondly, the only permitted insecticides were old ones (particularly lead arsenate, which is a stomach poison, and therefore the leaf surface is not toxic to beneficial insects). Thirdly, the first two measures made it possible to re-introduce beneficial insects. The result of this package was that yields of cotton began to rise dramatically in the late 1950s.

Spotted alfalfa aphid in California

The spotted alfalfa aphid (*Therioaphis trifolii*) was first introduced from Europe to California in 1954. By the late 1950s the aphid had developed resistance to organophosphate insecticides, and crop losses became critical. The courageous step was taken of applying an organophosphate (demeton) not at an increased dose, as is usual when resistance problems appear, but at a reduced dose! Probably less than 20% of the aphids were killed by this dose. Most survived, but so did many of the natural enemies. Although previously these had not been effective controls, they were now at a more favourable ratio to the surviving aphids and were able to exert control. The local natural

Fig. 13.1. Mixed cultivation in the Cañete Valley of Peru: part of a classic integrated control programme against cotton pests (courtesy of John Deere).

enemies were also re-inforced by the importation of three additional species of parasitoids. Within a year of applying this programme, the crisis was over. As a cultural measure, strip-harvesting of the lucerne was introduced in order to maintain some aphids and natural enemies on newly cut strips when older ones needed some insecticide treatment. Later on, varieties resistant to the alfalfa aphid were introduced, but the solution of the 1950s insecticide crisis with lucerne was found before such varieties became available.

Peach potato aphid on chrysanthemums in glasshouses in the UK

The discovery of lighting treatments to produce flowering pot chrysanthemums at any time of the year led to the concept of all-year-round chrysanthemums as a continuous production line in large glasshouses in southern England. The peach potato aphid (*Myzus persicae*) now had plants at all stages of growth simultaneously available for continuous breeding (Fig. 13.2). Under these changed conditions, resistance to the organophosphate insecticides used at the time was inevitable and developed rapidly. By the early 1960s, this

Fig. 13.2. The aphid *Myzus persicae* on all-year-round chrysanthemum (courtesy of the late F. Baranyovits).

resistance was 4000-fold. At first, the solution was thought to be biological control, but attempts to use aphid parasitoids proved unsuccessful (though today they are sold commercially to glasshouse growers). However, it was realized that, if broad-spectrum insecticides were to be avoided, another control for glasshouse whitefly (*Trialeurodes vaporariorum*), red spider mite (*Tetranychus telarius*) and thrips (Thysanoptera) would be essential.

To control whitefly biologically, the use of the hymenopteran parasitoid *Encarsia formosa* was resurrected from the 1930s (see Chapter 7). A biological control for red spider mite was already being developed in The Netherlands, employing a predatory mite (*Phytoseiulus persimilis*) from South America (Chile). This biological control of red spider mite was quite complex to operate. The predatory mite was in fact so voracious that chrysanthemum houses needed to be stoked with pest mites after a period following the introduction of the predator to maintain it and prevent it dying out. Some growers actually reserved particular houses for rearing the pest in order to keep the predator in the commercial houses supplied with food. The situation got even more bizarre when the predator became a problem because it was easily accidentally transported into the pest mite-rearing houses. The search was on for a pesticide one had felt would never be needed, a selective pesticide which killed only the biological control agent but left the pest alive! Dichlorvos (as Vapona strips hung in the pest-rearing houses) eventually proved effective, though the chemical has recently been banned in the UK.

The control of thrips posed the problem that no suitable natural enemies were known. However, several of the important species of thrips bury in the soil around the plants before emerging as adults, and so a soil drench of insecticide could be used to control a large proportion of the thrips population without interfering with biological control agents on the leaves. However, the original problem of the aphids remained. Just then, the carbamate insecticide pirimicarb (see Chapter 3), which is biochemically selective for aphids and so was originally rejected by industry as not being sufficiently broad-spectrum, was marketed. A programme combining biological control of red spider mite and whitefly with insecticidal control of the aphid and thrips was then successfully introduced for chrysanthemums and also cucumbers.

The integrated control concept

This book will, it is hoped, have made it clear that the components of pest and vector control which are most likely to be generally applicable are chemical control, biological control, host resistance and environmental/cultural control. In relation to these components, the three examples just quoted have one striking common feature. In all three solutions two particular components, chemical control and biological control, were used together. Thus Rachel Carson's 'fork of the roads' position (see Chapter 5) was in no way vindicated. In the event, the proponents of chemical and biological control, probably to the surprise of both, came together where the two converging, not diverging, roads finally met in the development of what was then termed 'integrated control' (see below).

Many other examples could be cited, particularly the successful resurrection of biological control in apple orchards of Nova Scotia over a 12-year period. Here broad-spectrum pesticides were replaced by a non-persistent plant extract insecticide (ryania) which allowed the egg parasites of moths (the major pests) to survive. Fungicide applications were largely restricted to glyodin, which had little effect on the arthropod fauna.

The examples of cotton, lucerne and chrysanthemum all reflect the original aim of 'integrated control' defined by Stern, Smith, van den Bosch and Hagen (1959) as 'applied pest control which combines and integrates biological and chemical control. Chemical control is used as necessary and in a manner which is least disruptive to biological control'. The latter sentence suggests that the chemical should be selective between the various life forms which might encounter it in the field, especially between the pest and natural enemies, including the natural enemies of other pest species.

In fact, Stern and his co-workers somewhat undersold integrated control when they regarded chemical control as being used in a way 'which is least disruptive to biological control'. The examples quoted earlier suggest that integrated control can go much further than this; it can even become pre-dominantly biological control made effective by using insecticides! In Peru, the insecticides which failed had only been introduced in the first place be-cause the beneficial insects were not giving adequate control on their own, and the use of a stomach poison as part of the solution to the problem enabled re-introduced biological control agents to survive. In California, some aphids were still killed with the low dose of insecticide and this made the difference between effective and non-effective biological control. In the chrysanthemum glasshouses, it was the introduction of the selective insecticide pirimicarb, which made it economically possible to contemplate biological control of red spider mite and whitefly.

Now pick up almost any other book on integrated control or pest manage-ment (or even biological control, come to that) and it will likely deal almost exclusively with agricultural pests. This is because the principal aim with medi-cal and veterinary vectors and pests has usually been to get as close as possible to eradicate such insects, not to manage them. The most effective tools for erad-ication have been chemicals. Nevertheless, integrated management has been practised, almost unknowingly, for many years. For example, during the first half of the twentieth century, malaria control was undertaken around centres of economic importance such as tea estates and mining camps to protect the workers. The approach was holistic, such as introducing fish into water tanks and wells, re-aligning water courses to eliminate shallow pockets of water, draining small marshes, using mosquito nets and administering quinine. At about the same time control of African sleeping sickness (trypanosomiasis) entailed removal of vegetation along rivers and roads which provided resting sites for the tsetse fly vectors, construction of game fences to exclude wild animals (reservoir hosts for the parasite), fly pickets to check traffic for hitch-hiking tsetse flies, and in parts of East Africa slaughter of game animals, and later insecticidal spraying. Another example of how several methods for vector control have been integrated is provided by the Onchocerciasis Control Programme in West Africa, which at the end of the twentieth century was dosing rivers with temephos and other insecticides, but also applying *Bacillus thuringiensis* var. *israelensis*, albeit forced to do so because of insecticide resistance problems. The control operation then became even more integrated when the drug ivermectin was administered to the human population to kill the nematode parasite causing the disease. This should surely be regarded as a modern example of integrated control, or management. In the more affluent areas, such as North America and Europe, there is increasing integration of insecticidal control of adult mosquitoes with environmental management of larval habitats, use of predatory fish and applications of microbial insecticides. The euphoria of disease eradication has been replaced by a more pragmatic approach, in which it is recognized that only in a very few instances might vector-borne diseases or medical and veterinary pests be eradicated. Usually the best that can be done is to manage diseases and vectors so that transmission is substantially reduced, and major pests become minor ones. However, progress in integrated control in less affluent countries is much slower, although there have been some advances in areas of China and India with regard to mosquito control and malaria.

Concepts of pest and vector management

As we have said, integrated management does not advocate a ban on pesticide usage, and not only allows, but even promotes their intelligent use in a way that is compatible with other methods, such as cultural, environmental and biological control. The aim is to minimize insecticide usage to reduce environmental contamination and damage to non-target organisms and supplement this more limited use of insecticides with other measures to reduce pest populations. This diversification of selection pressure on the pests will postpone as far as possible the appearance of resistance, i.e. slow down the so-called 'insecticide resistance race' (Chapter 5). But ecological damage is not caused just by insecticides. Draining marshes to eliminate sites of mosquito breeding and removing vegetation to eradicate tsetse flies can cause ecological problems, as can poorly researched or careless biological control introductions, such as the cane toad, *Bufo marinus*, in Australia (see Chapter 7). Often permanent drastic ecological changes can be caused.

An integrated approach usually means combining insecticidal and non-insecticidal control measures such as the use of predators and intermittent irrigation of rice fields (to control mosquitoes) or allowing plant resistance to slow the increase of a pest population so that biological control can have a greater impact. However, common sense dictates that there can also be integration of two different insecticidal measures, such as the application of larvicides to garbage dumps and use of dichlorvos strips in houses to kill house flies. Integrated management therefore implies the best combination of two or more measures. Although environmental measures such as physically preventing mosquitoes, such as *Aedes aegypti*, breeding in domestic water containers are commendable, if despite this there is a dengue epidemic threatening peoples' lives, then the only rapid response is insecticidal spraying to kill infected adult mosquitoes. However, once the epidemic has been curtailed or prevented then there can be a return to a more integrated approach to control the vectors. Similarly, in 2001 there was an introduction to California from Florida of a new leafhopper (*Homalodisca coagulata*) capable of wiping out the wine industry (by transmitting the devastating bacterium causing Pierce's disease). The immediate response has had to be the 'first aid' approach of insecticides, but the local scientists have immediately embarked on the search for a less pesticide-based package of control measures, especially since both the disease and vector have a huge plant host-range, including many crops but also many garden plants.

Although control efforts in crops are increasingly aimed at the pest problems as a whole, the concept of integrated control just described is still usually targeted at individual pest species, and therefore the stages in the practical implementation of integrated control are worth elaborating in a little more detail.

The procedure of integrated control

Establishing economic thresholds in agriculture

Economic thresholds are not a concept relevant to human disease vectors, because just single individuals are a threat to human health, a commodity on which mankind is reluctant to put an economic value. One 2002 attempt to do so comes up with about US$ 2.2 million as the value of a human life, since that is apparently the average amount a spouse would get if her policeman husband were slain in New York. However, for human disease vectors there is an analogous concept of 'critical thresholds', a discussion of which follows this section.

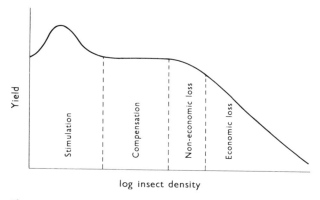

Fig. 13.3.   Generalized relationship between crop yield and pest density.

In crops, economic thresholds quantify the important decision of how far a particular pest population can be allowed to grow before insecticides must be applied to control crop loss. Insecticide applications can then be restricted to treatments which are as selective as possible and applied only when absolutely necessary. Stern *et al.* (1959) defined the 'economic threshold' (ET) as 'the density at which control measures should be determined to prevent an increasing pest population from reaching the economic injury level' (EIL). This is defined as 'the lowest population level that will cause economic damage'. The ET is therefore a threshold for action, related by experience and/or experimentation to the EIL.

The literature suggests that curves of yield reduction plotted against pest density can take many different forms, and these different forms of curve have often been related to different types of crop. However, to a large extent these different curves merely represent different parts of the generalized relationship illustrated in Fig. 13.3.

Low pest infestations may actually be beneficial to yield by stimulating plant growth through inducing the mobilization by the plant of wound-repair hormones or by allowing fewer fruits to develop to a greater size (thus reducing the need for chemical fruit thinners). Some crops (e.g. soya beans) are so leafy that many of the leaves are heavily shaded and so, although like all leaves they use up carbohydrate by respiration, they hardly contribute to photosynthesis and so can be a drain on yield. Indeed, one of the first improvements a plant breeder tries to make with a crop is to maximize yield by seeking the genetic trait which switches off further vegetative growth at the optimum leaf area (see 'leaf area index' later). Defoliation by hand or by insects of a proportion of the leaves on plants which have leaf area greatly in excess of this optimum can lead to an increase in yield.

After the phase of stimulation comes the phase of 'compensation'. This is a phase where increasing pest density does not reduce yield, since the plant can compensate in some way for damage. Cotton is a good example of a crop in which such compensation occurs. Provided the cotton plant is well fed and watered, a good crop will develop from 12 of the perhaps 90 flower buds that the plant produces. If none of these 90 buds is destroyed by attack of *Lygus* bugs or bollworm (*Helicoverpa armigera*), the extra buds will be shed naturally anyway. Crops such as cereals are often sown so densely that adjacent plants are in competition for light and nutrients. The result is that such crops will give an identical yield per hectare over quite an extensive plateau of plant density. Thus, provided the pest does not destroy cereal plants in patches, adjacent plants will take up the space of killed-out plants, and the larger surviving wheat plants then individually show an increased yield and compensate for the lost plants. Leafiness of crops is often responsible for compensation, especially for leaf damage. As pointed out above, plants give optimum yield at a particular ratio of leaf area to ground area (the 'leaf area index'). Thus damage to leaf area which is in excess of this ratio is unlikely to affect yield. Crop improvement by plant breeding is usually carried out under full insecticide protection, with no leaf area destroyed by pests.

Also as pointed out above, in seeking varieties with the highest potential yield, breeders select those that stop producing new leaf area after they have achieved the optimum leaf area index. As a result, the more improved that major world crops have become as a result of plant breeding programmes, the more has the capacity for compensation been bred out of the new varieties. A good example are many of the legumes, which have the potential to continue flowering and podding until they have set seeds, as long as water and nutrients are available. Quite heavy losses of flowers and young pods can therefore be compensated, albeit with a delay in the harvest. Plant breeders have put emphasis on including genes which switch the plant from vegetative growth to flowering at a much lower leaf area index, but also fix the number of flowers. The large potential for compensation found in the earlier varieties has thus been lost.

Once damage exceeds the compensating powers of the crop, further increase in pest populations will cause a progressive reduction in yield. However, as long as the damage would not be economic to control, pest levels still lie below the EIL. The costs that need to be taken into account are those of pesticide, labour, machinery and fuel as well as damage that pesticide application normally causes to the crop (machinery damage, the effects of soil compaction under the tractor wheels and often a measurable check to plant growth resulting from the toxin applied). As long as these costs exceed the damage caused by the pest the cost–potential benefit ratio remains above 1 and the EIL has not been reached. Potential benefit is, of course, not a fixed value for a given crop. It fluctuates with the environmental attitude of the grower concerned, his personal as well as local economic conditions, the state of the market for the crop, the cost of distributing the crop, the investment the crop represents and many other considerations. For example, 'organic' growers can obtain a premium for their produce as many consumers prefer eating food free of synthetic pesticides. The introduction of biological control with a pathogen of aphids on chrysanthemums had the surprising economic impact that the cut flowers which had not been treated with insecticide could be sold at a premium because of the more attractive soft appearance of the leaves.

In spite of all these variations affecting the relationship of the ET to the EIL, thresholds are increasingly used in practice. They can be established sufficiently accurately to guide the decision 'to spray or not to spray' once the relationship of yield to pest density has been defined for the crop. It is probably also true that thoroughly accurate ETs are not as necessary as was once thought. The real progress that can be made is to educate the farmer away from insurance spraying, so that pesticides are applied only to a challenge from the pest. Whether or not the farmer sometimes then sprays when it is not really necessary is perhaps not as important as that the frequency of spraying the crop is reduced. Farmers are certainly going to lose confidence in basing their spray decisions on economic thresholds if they result in failure to apply a spray when experience later shows that a yield loss was suffered. In contrast, farmers are not likely to discover that a spray they applied had not actually been necessary.

Critical thresholds in medical and veterinary entomology

The 'cost–benefit' concept, on which ETs in agriculture are based, is also applied for vectors of human and animal diseases, but with nothing to sell it is viewed somewhat differently. A minimum number of vectors, called the 'critical threshold' or 'critical density', is required to sustain transmission (see p. 284). Controlling the vector population to below this threshold can lead to disease eradication, or at least substantially reducing it to give disease control.

It may also be difficult to partition costs. For example, water management in rice fields may reduce mortality and morbidity caused by malaria, Japanese encephalitis and schistosomiasis (the intermediate hosts of which are aquatic snails). Therefore costing a reduction in a specific disease is not easy. Increased productivity by workers is sometimes treated as an indicator of improved health, but some dislike this as the emphasis is on people as producers. Moreover, only rarely have the reasons for days of lost productivity been carefully investigated, and even then the estimate may be unrealistic because of the ways in which families adapt to ill health. Improved health may extend life expectancy but if it does not extend working life this has little economic benefit, if that is to be the yardstick rather than human health per se.

There have been relatively few monetary estimates of losses caused by vector-borne disease to man, probably because they are difficult to make with any degree of reliability. In cost–benefit analysis, estimates are required of the costs of implementing control and the benefits that result. This can be difficult, because there are often 'hidden' costs of control such as training and networking and also because, as pointed out above, benefits of improved health are difficult to cost. It is easier with livestock, because improved animal health can be directly translated into monetary benefits in selling the animals or their products. Nevertheless cost–benefit analysis helps identify the economic burden vectors can impose. For example, malaria increases absenteeism from work and puts a strain on health resources to treat the sick (see below). Annual economic losses due to malaria have been reported at various times as US$ 23 million for Indonesia (1958), US$ 200 million for India (1942), US$ 54 million for Paraguay (1957) and US$ 50 million for Mexico (1955).

Cost-effectiveness analysis is similar to cost–benefit analysis, except that it is used to compare the monetary costs of different control approaches to achieve the same target, so that the most appropriate choice can be made:

- In 1974–97, the total cost of the West African Onchocerciasis Control Programme was about US$ 513 million, which equates to less than US$ 1 per person a year to protect against river blindness. Cost of APOC (African Programme for Onchocerciasis Control) (see p. 269) over a 12-year period is estimated as going to be US$ 161 million.
- Successful eradication of the New World screwworm fly (*Cochliomyia hominivorax*) from North Africa by release of sterile male flies cost about US$ 80 million over four years (1988–92). The benefit was valued at US$ 230 million. However, if this fly had managed to enter sub-Saharan Africa the livestock industry would have been devastated and the financial costs would have been enormous. The direct burden of screwworm damage to the American livestock industry was estimated as US$ 715 million in 1992, with an annual cost of US$ 35 million to keep screwworms from re-invading the USA.

- In 1999/2000 the days of disability per year per person in malarious areas ranged from 5 to 20. This represents an enormous economic burden. Cost of malaria control based on spraying houses with deltamethrin varied from US$ 2.6 per person per year in South Africa to about US$ 7.4 per person in Colombia compared with an annual cost per person of US$ 1.5 and US$ 3.7 per year when pyrethroid-impregnated mosquito nets were used. Clearly either control strategy is cost effective, but impregnated nets especially so.
- In 1991, to protect a family in Paraguay against Chagas disease cost only US$ 29 if their house was sprayed with lambdacyhalothrin, compared with as much as US$ 700 if there were improvements to the house. Insecticidal spraying is thus the most cost-effective measure despite it having to be repeated about yearly.

How are thresholds established?

Techniques are available for determining the economic injury level for crop pests, i.e. the lowest population likely to cause damage in excess of the costs of control (see above). However, this then has to be translated to the economic threshold, when the farmer needs to take action. This translation is not so much a matter of science as of judging the time it takes for the farmer to apply the control measure in relation to the potential increase rate of the pest population.

Surveys attempt to correlate yield losses with pest density by recording both over a wide range of the latter. As this often involves surveying a large geographical area where the crop is grown, the very fact that widely differing pest densities can be recorded is itself a warning that correlation with yield may be spurious. It is likely that yield may be directly affected by other factors which happen to be correlated with pest density, e.g. longitude, latitude, altitude, soil type, local intensity of plant diseases or even choice of crop variety typical for an area. Survey data need close inspection to check for such internal correlations; apart from this, surveys do provide an easy route to establishing economic injury levels.

The obvious approach is to infest plants to known infestation levels, and then to measure yield. This is unfortunately extremely difficult in practice, certainly on a field scale. Many insects are mobile, and tend to move around so as to cancel out initial differences in density. Cages may be needed, but these can change the physiology of the plants grown within them and, through this, pest reproduction rates. Moreover, where the insects are fast-breeding, e.g. aphids, they will complete several generations before effects on plant yield can be measured. During this multiplication, the contest competition effects described in Chapter 2 will effect a reduction in breeding rate correlated with pest density, so that the initially smaller populations begin to catch up with the larger ones. With such effects, it is impossible to relate initial infestation to the population which actually caused damage.

Insecticides provide the possibility to manipulate pest densities experimentally. Many insecticides unfortunately have their own direct effects on crop yield, but this can be catered for with suitable controls. Insecticides are certainly useful for checking a rising population at different points in time and seeing which is the latest application which satisfactorily protects yield.

Given the problems of establishing EILs by techniques which use the insects themselves, many workers have sought to simulate pest damage by methods such as using a pair of scissors to create levels of defoliation. Results of simulation experiments usually suggest that the plant can tolerate much more damage than is suggested by equivalent experiments with insects. One important reason for this is that insects defoliate bite by bite, and therefore cause far more wounding to the plant than a single cut with a pair of scissors. Another very important reason is that insects do not just feed but also add saliva, which itself can severely damage plants, sometimes more than by the removal of plant tissue.

An interesting technique, which might be termed 'grower bioassay', was used in Kentish apple orchards to get directly at some kind of useful ET. Over several seasons, advisory officers and growers counted pests on the trees and progressively delayed the pest level at which treatments were applied until the grower was convinced that economic loss was being suffered. This technique may not have established scientifically valid thresholds, but it certainly gave the perhaps rather more useful information of what was a 'grower worry threshold'.

Although we may speak of the goal of achieving 'critical thresholds' or breakpoints for signalling when vectors of diseases to man and livestock need controlling, this approach is not widely used. Moreover, we usually have little idea what the values are, in comparison with pest thresholds on crops, however imprecise the latter may be. Nevertheless, there have been some successes in identifying critical numbers, such as the number of mosquitoes caught in light-traps below which there is no transmission of an arboviral disease, such as western equine encephalitis.

In summary, critical thresholds have as yet not proven to be of much practical value in the planning of vector control strategies in the way economic thresholds now form the basis of such strategies in agriculture.

How are thresholds used in practice?

A direct and commonly used approach is to count the pests on a sufficiently representative number of plants in the crop, a procedure known in many parts of the world as 'scouting'. For example, the ET for the grain aphid (*Sitobion avenae*) in England from the second node detectable stage to early flowering is 50% of the plants infested with aphids, monitored weekly.

If predators are numerous, the farmer should wait until early flowering before further monitoring. From early flowering to the milky ripe stage, the threshold is an average of five aphids per ear, or 30 aphids on the flag leaf, on a sample of 20 tillers, monitoring every field every two days. Once the crop has reached the milky rips stage, any control would be uneconomic. Farmers should monitor crop areas well away from the field edges, where unrepresentative higher populations often occur. Such crop monitoring operated by the farmer is time-consuming and asking rather a lot!

In fact, the EIL from early flowering onwards is nearer 20 aphids per ear, and this example illustrates the loose relationship between the EIL and the ET. Not only does an ET of five give time for the farmer to take action, but also it is accepted that the field counting is likely to underestimate the true number of aphids (the small younger ones are easily missed). The farmer is more likely to use thresholds if the trouble of counting individual pests can be avoided. So it is fortunate that, for black bean aphids, the economic threshold can be stated more simply; it is when 5% of the stems are infested, and other economic thresholds are based on simple measures such as per cent defoliation rather than a count of the number of caterpillars.

Monitoring becomes even simpler and therefore attractive to the farmer if the counting can be restricted to relatively small numbers of pests in one or a few insect traps. These also usually avoid the need for anyone to have to walk into the crop. Traps are therefore increasingly being used for monitoring, and those emitting sex pheromones (Chapter 10) have proved particularly valuable as shown by their wide uptake. However, it cannot be assumed that research can produce such a pheromone trap monitoring system for any pest. As emphasized in Chapter 10, it is the female pests which emit long-range sex pheromones; these pheromones (whether natural or synthesized) will only trap males. With many pests, there is no good correlation between the number of males caught at traps releasing female sex pheromones and the number of eggs subsequently laid on the crop by females. Some examples of the use of pheromone traps for monitoring crop pests in Britain, and the thresholds used, are given in Chapter 10, and will not be repeated here.

Fig. 13.4.   A discarded tin can being used as an ovitrap to monitor changes in relative population size of *Aedes aegypti* (M. W. Service).

Monitoring for vectors of human and animal diseases can take many forms and incorporate a wide variety of sampling techniques to monitor relative changes in population size of pests and vectors. A successful programme needs to be cheap and simple.

Aedes aegypti is a vector of yellow fever in Africa and South America and of dengue fevers in many warm countries. Larvae occur in domestic water pots (Fig. 2.7) and other water-filled discarded utensils found scattered around human habitations (Fig. 13.4). Several indices have been used to measure larval densities, such as the number of receptacles with water containing larvae

Fig. 13.5. A battery-operated light-trap placed in a bedroom in a Guatemalan village for monitoring population levels of mosquitoes such as anophelines (M. W. Service).

(container index), or the total number of receptacles with *Aedes aegypti* larvae per 100 houses (Breteau index). Often a Breteau index above 5, or a container index greater than 20, indicates that the population of the vector has reached a level which presents a threat of urban transmission of yellow fever. However, the validity of these warnings is debatable, if not for any other reason than a low incidence of yellow fever virus in the population or in the monkey population (reservoir hosts).

Some mosquito species of arboviruses, such as *Culex tritaeniorhynchus* (vector in Asia of Japanese encephalitis) and *Culex tarsalis* (vector in USA of western equine encephalitis) can be caught in battery-operated light-traps (Fig. 13.5), sometimes supplemented with dry ice for production of the attractant carbon dioxide. When the numbers caught reach a predetermined index this signals that control operations should be instigated to prevent the threat of disease transmission.

In the Onchocerciasis Control Programme in West Africa the early detection of any upsurge in vector population caused by control failures relied almost entirely on using people to attract and catch the simuliid black fly vectors, in so-called human bait catches or, as the WHO euphemistically prefers to call them, landing catches. The reader will appreciate there are ethical considerations in using people in this situation, but there are no suitable alternative methods for monitoring man-biting simuliid black flies.

Although ticks are vectors of infections of humans, economically they are more important in spreading animal infections, especially parasites of cattle. The presence of ticks can be monitored by introducing one or two cattle in pastures and periodically counting the ticks parasitizing them, or by counting the numbers of ticks caught by blanket-dragging. In this technique, ticks cling to a sheet of denim or other coarsely woven cloth that is slowly dragged across the ground. Alternatively ground-based traps containing dry ice as an attractant can be used. Ticks entering the trap are caught on sticky tapes. High tick densities will indicate that control measures are merited.

Monitoring diseases rather than the vectors can be passive or active. Passive monitoring is based on more or less chance encounters with infected animals or reporting infected humans diagnosed at their homes or at clinics or hospitals. With active monitoring detection surveys are undertaken to detect infections in animals or humans. This is the more informative type of surveillance.

Another approach is to monitor current epidemiological data to detect the possible beginnings of an epidemic. With malaria this can comprise plotting the monthly mean numbers of malarial patients over the last five years together with an estimate of the upper limit of the normal range, such as the 75th percentile, over these years, and then comparing this with current data. When these figures exceed the five-year normal range then an epidemic is likely and control measures can be considered. But frequently, especially in remote rural tropical areas, there is no warning and an epidemic may become established before it is detected and consequently control operations are late in starting.

The threat of barley yellow dwarf virus spreading from grasses to cereals in the autumn is based on using molecular methods to assess the proportion of aphids (*Rhopalosiphum padi*) caught in suction traps in the autumn which are infected with the virus. Analogous is the estimation of human or livestock disease outbreaks by the collection of potential vectors and determining whether or not they are infected with pathogens that will cause diseases, such as certain viruses in mosquitoes and ticks. An even better, and usually easier, method is to place 'sentinel' animals, usually mammals or chickens, in enclosures or cages and to bleed them at regular intervals to see whether they have become infected with disease agents, due to being bitten by infective vectors. This procedure is most commonly used in the study of arboviruses transmitted by arthropods.

Forecasting pest and vector outbreaks

Agricultural pests

Forecasting is a further stage towards predicting the need to spray without the grower personally being involved in monitoring or predicting the most effective spray date given the presence of an economic threshold in pest numbers. For pea moth (*Cydia nigricana*), climate is the most important variable. When the grower finds moths in the pheromone traps have reached the threshold (ten moths in either of two traps at right angles, and in two consecutive weeks) a phone call to an Agricultural Development and Advisory Service computer provides an estimated spray date. This date, estimated to coincide with the hatching of the larvae, can be many days in the future. The computer program includes the normal delay between the appearance of males in the traps and the females laying their eggs, plus the delay between the date of oviposition and hatching of the eggs. This latter part of the prediction calculates the development time in the egg as a function of the forecast daily temperatures but is updated as actual temperature records are taken. Growers can then refer back to the Advisory Service computer for a more accurate prediction as the provisional spray date approaches.

The development of pea moth eggs is calculated from temperature records of 'day degrees'. Each part of the life history of an insect will only proceed above a certain temperature (the development threshold). Every degree the mean temperature for a day is above this contributes one 'day degree' to a temperature sum. When this sum reaches a certain total, development is predicted to have been completed. This technique of accumulating day degrees has been used for a long time to predict the appearance of pests on the crop after winter. For example, spraying for the mealybug *Pseudococcus* in apple orchards in Japan is recommended when 220 day-degrees above a threshold temperature of 9 °C have accumulated.

A novel approach to the use of temperature sums is 'Tempest', a little electronic temperature accumulator available from Insect Investigations, Cardiff in the UK. Different models of Tempest are available for a number of important pests. The device is placed in the field and shows a green window until the temperature sum at which it is the correct time to spray is reached.

Increasingly, computer modelling of plant growth and weather forecasts updated by current weather conditions are being developed to try to produce long-range forecasts from early monitoring data. Some quite long-range predictions relate to the effects of severe winters on pest populations. Thus the annual variation in the incidence of beet yellows virus on sugar beet, vectored by the aphid *Myzus persicae*, can be predicted by calculating the number of days with frost in the previous January to March and how far the following April temperatures deviate from normal. Another example relates to cutworm (caterpillars of noctuid moths in the genus *Agrotis*). High attack of vegetable crops by cutworms is correlated with high survival of the caterpillars to the third instar. This survival can be modelled from temperature records, corrected for rainfall on the prediction that each 0.1 mm of rain reduces the number of first and second instar caterpillars by 1%.

A simulation model developed at Horticulture Research International (HRI) in the UK uses temperature to forecast the appearance of the generations of cabbage root fly (*Delia radicum*), though not the size of the attacking population, since the flies are very mobile and move large distances. There is a normal distribution covering a wide range in development time of the insect, largely produced by the existence of an 'early' and a 'late' biotype in terms of emergence after winter diapause. The model therefore simulates the development of an imaginary cohort of 500 individual insects. The forecast is being used by commercial growers.

Egg sampling is used as a basis for a long-range forecast of the future crop damage that can be expected from some pests. For many years the Agricultural Development and Advisory Service has monitored egg populations of wheat bulb fly (*Delia coarctata*) in fallow fields during August and September. A threshold of 2.5 million eggs per hectare has been established as predicting that spraying will be necessary when the eggs hatch early in the following year, although sometimes high egg counts are not followed by serious damage since the eggs are subject to depredation from ground-dwelling carabid beetles. From 1970, the Advisory Service was involved in counting the eggs of the black bean aphid (*Aphis fabae*) during the winter on selected spindle bushes as a basis for separate long-term forecasts of attack of field beans by the aphid for 16 areas in southern England. This enabled successful predictions about where chemical treatment would be unnecessary or clearly necessary; between 1970 and 1982 only 10% of the total bean area involved received a 'damage possible' forecast, which then required additional shorter term crop monitoring. With a parallel forecast now available for different areas from the network of suction traps operated in the UK by the Rothamsted Insect Survey, the laborious sampling of eggs on spindle bushes has been discontinued.

Fig. 13.6. The 12.2-metre high suction trap used by the Rothamsted Insect Survey (courtesy of Rothamsted Research).

This network of suction traps is one of the considerable efforts being devoted to developing forecasts based on evaluating the dispersing populations of some pest species. The network consists of 15 suction traps (Fig. 13.6) distributed over the UK and over 40 elsewhere in Europe sampling flying insects at a height of 12.2 m. Particularly migrant aphids have now been monitored for over 30 years and, for some species additional to black bean aphid, forecasts have been developed to predict both the likely size of the crop infestation and its timing. In collaboration with other European scientists, the Rothamsted team is working to develop accurate forecasts for cereal aphids; especially those involved in bringing barley yellow dwarf virus to the crop. Forecasts provided for the farmer, and not involving any time-consuming input on the farm, are the ultimate goal.

Medical and veterinary pests and vectors

Forecasting outbreaks of medical and veterinary pests and vectors has not been pursued as much as for agricultural pests, although more recently satellite imagery has been employed (see p. 293). There are, however, a few predictive signs that can signal vector problems, some examples of which are given below.

Weather monitoring.   Temperature and precipitation are the key climatic elements used in predicting pest or disease outbreaks. For example, increased temperatures may shorten vector life-cycles and give rise to increased pest populations, or extend their range into areas that were formerly too cold for their survival or for disease transmission. In mountainous regions the depth of snowpack has been used to predict outbreaks of so-called 'snow-melt' mosquitoes, i.e. *Aedes* species that will breed in water pools when the snow has melted in the spring. Using this criterion, advance warning can be given months ahead, but because there are so many other environmental factors, as well as flood-control measures, the predictive value is not always very reliable. Nevertheless floods and excessive rainfall can sometimes give several weeks or more warning that mosquito populations might reach abnormally high densities.

Attempts have been made to develop predictive models. For example, in 1997 a model incorporating more than 180 epidemiological and environmental parameters was developed to predict outbreaks of Lyme disease (caused by the spirochaete *Borrelia burgdorferi*), an infection spread by ticks throughout much of the world. This programme is still being evaluated.

Disasters.   Disasters, such as earthquakes, hurricanes and war are often the precursors of pest and vector outbreaks and disease epidemics. For example, body lice (*Pediculus humanus*) proliferate in refugee camps and raise the risk of louse-borne typhus (*Rickettsia prowazekii*), and accumulation of polluted waters provides breeding places for the mosquito vector (*Culex quinquefasciatus*) of bancroftian filariasis. Preparedness is essential if disease epidemics are to be prevented.

Similarly, much can be done to prevent pest and vector problems arising in development projects, e.g. dam construction and irrigation schemes. But good knowledge of the pest problems likely to be associated with a project is essential, and methods to minimize, or even prevent, these problems arising need to be implemented at the planning stage.

Increasing use is being made of computerized geographical information systems (GIS) by which data are graphically displayed in multilayer maps. For example, in California, data generated by the Californian Vector-borne Disease Surveillance System are posted weekly on a World Wide Web site. In future there will be greater use of this means of communication, more accurate interpretation of such data, and more reliable predictions of pest problems.

Satellite imagery.   Although in the 1960s we were able to see photographs of the earth from space, the first civilian satellites were launched by the USA in 1972. In 1975, Landsat satellites pioneered the way the earth can be viewed and generated for the first time high-resolution imagery. These satellites are maintained in polar orbit at a height of about 900 km and scan a swath of 185 km with a repeat cycle every 16 days. Other commercial satellites cover a width of 2000 km every 12 hours or 60-km swath every 26 days. Selection of satellites for imagery depends on the type of information required.

Many studies involving disease vectors have focused on remote sensing to map potential vector habitats based on detection of rainfall, saturation deficit, water, types of vegetation and soil. However, identification of potential habitats does not necessarily mean that they actually support vector populations. This limitation has led to criticisms that the value of satellite imagery is overrated and interpretation of habitat maps is at best dubious. For operational surveillance and vector control programmes it is necessary to make predictions, based on habitat mapping, of the risks of pest or disease outbreaks; a few studies have attempted this. For example, there have been studies to establish the degree of risk of malaria transmission based on the proximity of houses to flooded areas considered suitable for anopheline breeding.

Other studies have taken the relationship between vegetation type, saturation deficit and population density of tsetse flies, and then used satellite imagery to identify areas with high potential for tsetse flies. As another example, the distribution of the tick vectors of Lyme disease depends on interaction of vertebrate hosts, humans and landscape patchiness such as areas of coniferous forests, deciduous forests, mixed forests, glades, recreational areas, housing development, tracks and roads. Remote sensing data have been used to identify and map landscape features, such as those above that are associated with transmission of Lyme disease.

What of the future? The potential of satellite imagery is still far from fully exploited, but studies on medical and veterinary pests and vectors have shown that, despite some of their current limitations, remote sensing and GIS technologies have much to offer, especially in gaining new perspectives of how landscape ecology influences disease transmission. One goal is building predictive models that can warn of potential pest and disease outbreaks, but for many scenarios greater sophistication will be needed for this.

There is also some initial interest in satellite imagery in relation to plant pests, as 'false colour' imaging can identify patches of forest damaged by defoliating insects and trigger spot-treatment of insecticides. It has even been suggested that someone sitting at a monitor thousands of miles away might be able to identify wild grassland areas where locust damage is occurring, and guide spray teams to these areas to prevent swarms developing.

Assessing potential natural enemy activity in crops

Naturally occurring biological control almost always provides an important component of pest management in crop ecosystems. Sampling is necessary to determine whether natural mortality agents are present in sufficient numbers to be worth conserving with selective chemical control. Techniques for doing this include comparing sprayed and unsprayed crops, sampling with various traps and vacuum samplers. Also, pests can be dissected to look for immature stages of parasitoids or returned to the laboratory to breed out the adults; immunological techniques using antibodies (e.g. ELISA) and other molecular techniques (e.g. DNA analysis) can be used to identify prey proteins inside the gut of potential natural enemies.

Augmenting the resistance of the agro-ecosystem

The purpose of augmentation may be to provide natural enemy action where this is insufficient (e.g. in heavily sprayed ecosystems, crops with an introduced pest) or to establish a new equilibrium pest population at an artificially low level (Fig. 13.7). Introduced pests can sometimes have their populations equilibrated at a lower level by the importation of natural enemies (especially parasitoids) to re-establish the stabilizing influences existing in the country of origin from which the pest has escaped. Where parasitoids have disappeared because a vital alternative host has been lost through monoculture, replacing a single plant species, e.g. blackberries (*Rubus fructicosa*) near vineyards in California (see p. 166), may be all that is necessary. The introduction of partial plant resistance to the pest will slow down the rate of pest increase, and the existing natural enemies may well then regulate the pest to a lower equilibrium

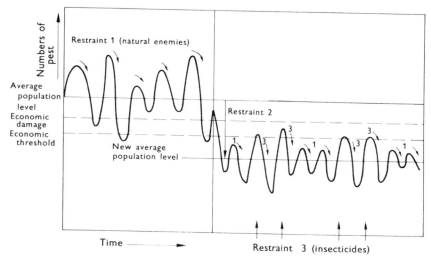

Fig. 13.7. Diagrammatic representation of the role of three restraints in an integrated control programme. Restraint 2 may be an imported natural enemy or a pest-resistant crop variety (From van Emden, 1969; courtesy of the Society of Chemical Industry).

level. In augmenting the resistance of the environment, cultural controls are also worth considering. Two particularly valuable types of measures are (a) those which make conditions more suitable for natural enemies, such as the provision of refugia or adult food such as nectar sources and (b) measures such as the destruction of crop residues, which may break the life cycle of the pest in the region so that numbers in subsequent generations are dependent on immigration from outside.

Developing selective pesticide applications

The biological control potential already present in the environment or suitably augmented then needs protection from the sprays that are necessary whenever pest populations reach the economic threshold. It would obviously be ideal if we could use chemicals which were inherently selective. However, few inherently selective compounds have been developed, or are likely to be. The economic problems of developing insecticides were discussed in Chapter 3; very few single crops, disease vectors or pest problems offer a sufficiently large sales potential to warrant the development of a specific insecticide tailored to be selective for a particular integrated control solution. Moreover, perhaps too

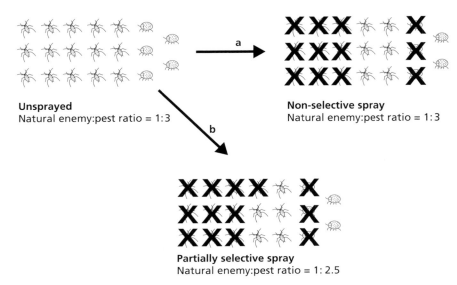

Unsprayed
Natural enemy:pest ratio = 1:3

Non-selective spray
Natural enemy:pest ratio = 1:3

Partially selective spray
Natural enemy:pest ratio = 1:2.5

Fig. 13.8. Cartoon of aphids and ladybirds to illustrate the importance of natural enemy : pest ratios. Spray (**a**) kills both pests and natural enemies to leave unchanged the ratio of 1 : 3 that existed before spray application. Spray (**b**), by killing just one more aphid in the cartoon, changes the ratio in favour of the natural enemies (from 1 : 3 to 1 : 2.5).

much emphasis has been placed in the past on selectivity between the target insect and its natural enemies. There are many problems (e.g. low-density pests such as vectors of plant or animal diseases) where the pest virtually has to be eliminated before there is a worthwhile reduction in disease transmission, or cosmetic crop damage is reduced to customer-acceptable levels. This can usually be achieved only with a pesticide (including one delivered through GM crops) or with sterile male insect release, and the natural enemy specific to the pest might almost as well be killed by the pesticide as allowed to die of starvation or emigrate because of the disappearance of its prey. In such cases, integrated control involves pesticide selectivity between the pest in question and the natural enemies of other potential pests in the same crop/ecosystem, so that insecticide control of the key pest does not lead to an upsurge of other pest problems.

Apart from the few at least partially selective insecticides which are available, it is often possible to find ways of making a broad-spectrum insecticide selective by the way we use it. When we spray, we will certainly kill both pests and natural enemies; the secret is to make sure that the pesticide application changes the natural enemy to pest ratio in favour of biological control (Fig. 13.8). We may

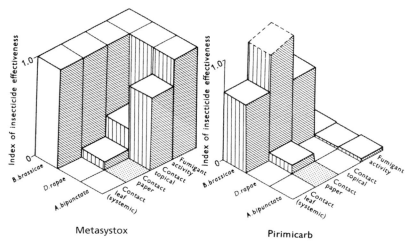

Fig. 13.9. Comparative toxicities of contact and fumigant activities of two systemic insecticides to the cabbage aphid (*Brevicoryne brassicae*), its main parasitoid (*Diaeretiella rapae*) and a common predator (the ladybird *Coccinella bipunctata*). Toxicities are expressed as proportions of the effectiveness of the systemic kill of the aphid by metasystox (courtesy of G. D. Dodd).

even be able to accept a slightly reduced kill of the pest by our application if, as a result, we succeed in shifting the balance in favour of biological control. Thus it is possible to use insecticides to help make biological control more effective. Various sources of selectivity in pesticide applications are discussed below.

Inherently selective insecticides
A few pesticides possess selectivity for biochemical reasons. Mention has already been made of the aphicide pirimicarb, a systemic and fumigant carbamate which affects aphids and Diptera, but not other taxa including ladybirds or aphid parasitoids (Fig. 13.9). Another widely used insecticide with some selectivity is the organochlorine endosulfan, which has been used to kill tsetse flies resting on vegetation, although spraying vegetation around water should be avoided because endosulfan is very toxic to fish. This compound seems fairly safe to insects in the order Hymenoptera, which includes most of the parasitoids used in biological control. We should also consider other materials such as insect pathogen sprays (see Chapter 8) and internal growth regulators (Chapter 3), which tend to be much more selective than traditional insecticides and are therefore more compatible with integrated control. For example,

Bacillus thuringiensis var. *israelensis* can safely be used to kill the aquatic larvae of mosquitoes and simuliid black flies because it is relatively non-toxic to other macro-invertebrates. There is increasing research on testing pesticides against natural enemies in the field for particular crop situations. Some selectivity can often be found in such field trials, even with insecticides which are known to be very broad-spectrum. For example, in Nigeria the carbamate methomyl gave good control of the legume pod borer (the pyralid moth *Maruca testulalis*) in cowpeas without affecting its major hymenopterous parasitoid (a species of *Cotesia*, Braconidae). Such partial selectivity revealed in field trials is extremely useful, though there may well be other reasons than biochemical selectivity of the insecticide for the observed effect. Selectivity may relate to other factors such as the behaviour of the organisms and how much contact they have with the pesticide. The message is that a little experimentation may show unsuspected partial insecticide selectivity.

Formulation of insecticides

Pesticides may be available in different spray formulations, such as emulsions or wettable powders (see Chapter 3). Trying different formulations may reveal differences in selectivity between them. Such differences again may be due to the behaviour of insects in relation to the type of deposit remaining on the plants, the persistence of different formulations and the contribution of the different additives in the formulation to the toxicity of the residue. Many insecticides can be sprayed in an encapsulated form to achieve selectivity. Droplets can be coated with polymer; the encapsulated droplets can be sprayed in water, or even as formulations which cause the droplets produced from the nozzle to form a capsule on exposure to air on the way to the target (see Chapter 3). Encapsulation thus converts the contained pesticide into a stomach poison, and natural enemies can move over the leaf safely, since the insecticide is specifically addressed to leaf-chewing insects.

Reduced dose rate

In describing the solution to the problem of resistant lucerne aphids (*Therioaphis trifolii*) in California (see earlier in this chapter), the crucial concept was that a reduction in dose rate would improve the natural enemy to aphid ratio. Why should this be so? The answer lies in the different type of response (Fig. 13.10a) shown by many carnivores in comparison with many herbivores. The range of insecticide dose spanning from low kill to high kill of herbivores is generally larger than that for carnivores. The reason for this is not entirely clear, but may well be connected with the wide range of enzymes, some of which can detoxify pesticides, which herbivores need to cope with the

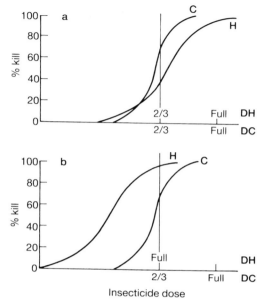

Fig. 13.10. Theoretical graphs of % kill plotted against insecticide dose for a herbivore (H) and a carnivore (C) **(a)** typical graphs; **(b)** graphs on a pest-resistant variety on which the insecticide dose needed to achieve the same kill of the herbivore as on the susceptible variety can be reduced by one-third (see p. 312). DH, dose scales relevant to the herbivore; DC, dose scales relevant to the carnivore (from van Emden, 1987; courtesy of Academic Press).

many secondary compounds they encounter in their host plants. Whatever the reason, the result is that the per cent kill of natural enemies decreases faster than that for pests as we decrease the insecticide dose. Selectivity of the insecticide thus increases until, at a very low dose, it may even be possible still to kill pests without killing any natural enemies at all. It is therefore clear that reduced dose rates may be an effective complementary mortality to that provided by biological control; yet the legislation in many countries against reduced dose rates, passed to prevent the insecticide manufacturers' recommended rates being exceeded, can make this generally applicable integrated control solution illegal.

Time of application
To obtain selectivity in time relies on knowledge of the life history of both the pests and the natural enemies. There may be times when a high proportion of the population of natural enemies is protected from contact with sprays. They

may be inside a protective casing (e.g. an insect egg or, for aphid parasites, a mummified aphid) or outside the treated area (e.g. preying on an alternative host in the hedgerow or outside the crop seeking flowers for adult feeding). Work in Nigeria has suggested that there may be potential for improvements in the ratio of natural enemies to aphids by avoiding applying insecticide early on before the natural enemies, which arrive in the crops well after the aphids, become abundant. Fewer aphids release less volatiles that attract natural enemies into aphid-infested crops (see p. 162) and also ladybirds lay eggs in proportion to the number of aphids they find. The suggestion has therefore been made that, provided aphid densities are not damaging, insecticides should be delayed until the ladybirds have been attracted and laid their eggs. An ephemeral spray at this time would reduce aphid numbers without killing the ladybirds developing within the egg, with the result that the hatching larvae would be in a very favourable ratio to the remaining aphids.

In Pennsylvania apple orchards, populations of spider mites (Tetranychidae) build up in between the two population peaks of the leaf roller moth *Phyllonorycter blancardella*. By carefully restricting spraying to the early part of June and the latter part of August/beginning of September, insecticides can be used effectively to control the leaf roller without damaging the ladybird populations which are important predators of the mites between these spraying dates.

It may even be possible to obtain some selectivity by spraying at a specific time of day. In both cotton and grain legumes, leafhoppers (Cicadellidae) (which normally are difficult targets for spraying since they feed on the undersides of the lower leaves) move up the plant in the evening and sit on the upper sides of the upper leaves. This makes them much easier targets for spraying, and they can be controlled with much less insecticide and much less penetration of the crop, giving the insecticide application considerable selectivity in favour of natural enemies.

Aerial spraying, often with non-residual insecticides such as endosulfan, for tsetse fly control can usually only be undertaken during temperature inversions, that is at dusk or dawn. But over flat terrain spraying can also be done at night, and this tends to reduce the kill of those non-target insects that are inactive at this time.

Application in space

Selectivity in space means treating only parts of the crop to enable natural enemies to survive in the untreated parts. A very simple and common way to achieve such selectivity when using systemic insecticides is to apply the toxin as a granule on the soil rather than as a spray to the leaves. The roots then take up the poison, but the leaf surface remains safe for natural enemies to move over. The modern technique of film-coating seeds with insecticide to protect them against soil pests (see Chapter 3) localizes the toxin and enables minute rates to be used; this must make such pesticide application almost fully selective.

In the example quoted earlier of lucerne in California (see p. 273), strip harvesting made it possible for the natural enemies to survive in more recently cut but re-grown strips of the crop whenever the adjacent taller strips required insecticide treatment. This concept of 'band' spraying, where only certain bands of the crop are treated on each occasion that insecticide is applied, can be used directly in field crops without complications such as strip harvesting.

There are a number of ways of implementing the band spraying philosophy in orchards also. Scale insects (Coccoidea) in citrus have been controlled by alternating the use of biological and chemical control on adjacent rows on a two-year schedule; i.e. alternate rows were sprayed with oil emulsion and left unsprayed in alternate years. Insecticide-treated attractive baits may be used to separate the pest from the natural enemies and 'lure it to its doom' (see p. 208), a simple technique that works well for house flies and cockroaches as well as for crop pests. A very nice combination of a variation on band spraying and the use of attractants has been tried on citrus in Australia against fruit flies (especially the Mediterranean fruit fly *Ceratitis capitata* and the Queensland fruit fly *Bactrocera tryoni*). In several countries, researchers have experimented with restricting spray to the lower half of the tree. Like band spraying, this achieves a fair control of the pest while leaving some of the tree relatively free of pesticide and enabling natural enemies to survive. The Australian workers, however, added an attractant (yeast hydrolysate) to their insecticide (i.e. they formulated a bait spray), which of course had the result of bringing many flies from the unsprayed upper part of the citrus tree down to the lower part to pick up insecticide there. A further development in some countries has been to separate the bait spray entirely from the tree, and apply it to any other available surface such as the ground vegetation. Somewhat similarly, in tsetse fly control only the bases of selected trees are sprayed with residual insecticides, which apart from cutting costs also lowers the kill of non-target insects and reduces environmental contamination.

Perhaps the most ingenious use of restricting pesticide application in space to improve the natural enemy to pest ratio is an example from coffee as early as the 1960s. One of the most intriguing aspects of this example is that DDT, the widely used broad-spectrum insecticide which did so much damage to biological control in the 1940s and 50s, was the agent used in this example to improve biological control! The main pest of coffee at that time in

Kenya, where the technique was used, was the giant looper (*Ascotis selenaria*). DDT was painted around the trunk of each tree in the plantation at the beginning of the season. Whenever looper caterpillar populations increased to an unacceptable level, the canopy of the tree was sprayed with a natural pyrethrum at a dose to achieve 'knockdown' of insects rather than kill them. A large number of insects thus fell out of the trees on to the ground; the caterpillars could only regain the tree by passing over the DDT band and thereby picked up a lethal dose of insecticide. By contrast, many of the natural enemies, not only of the caterpillars but also of other pests of coffee, regained the tree by flying upwards and thus missed the DDT band altogether.

The fate of the integrated control concept in respect of crops

Fate of the early examples of integrated control

It is important to realize that, although the early introductions of integrated control in Peru, California and British glasshouses described earlier were highly successful in providing solutions to the insecticide-induced crises of those times, the control systems adopted did not have long-term viability. The Cañete Valley in Peru has now largely returned to a monoculture of cotton, and strip-cutting of lucerne and the use of low-dose insecticide is no longer practised in California. Moreover, although many growers still use some biological control in British glasshouses (e.g. for cucumbers), the package developed for year-round chrysanthemums is hardly practised today. The reason for all this is the advent of the synthetic pyrethroid insecticides which, at least for a time, are proving effective at controlling the pest problems created by resistance to the earlier insecticides. Thus growers have been able to return to the pest control solution they find preferable, namely routine reliance on insecticides. This is because it is instant, they have control over it and the pesticide manufacturers will compensate growers if their products fail in spite of compliance with the instructions. However, insecticide-induced crises are going to recur and are indeed presently appearing in many places around the world, so that work and thinking about integrated control is still needed. This approach applies to both agricultural as well as medical and veterinary pests.

In tropical countries (where heavy insecticide use is often uneconomical and pest reproduction rates are very rapid), as more insecticides fail as a result of resistance and as public pressure for reduction in pesticide use escalates, the ideas of integrated control will certainly increase in importance.

Later developments of the integrated control concept: crop pests

Already by the mid 1960s, the original concept of an integration of chemical and biological control had been extended by H. H. de Fluiter (The Netherlands) and J. M. Franz (Germany) to embrace all suitable pest control methods integrated in a compatible manner ('harmonious control'). Harmonious control had the merit that cultural methods were more closely examined for their control potential apart from their effect on natural enemy abundance. Unfortunately, it also introduced the danger of trying to avoid chemical control altogether rather than stressing the better use of chemicals which was the cornerstone of integrated control. In 1967, at an FAO meeting, the Californians redefined 'integrated control' in terms very similar to harmonious control. In so doing, integrated control was relegated from something immediately practical (the ingenious use of pesticides) to theorizing from the armchair on what might be possible in the future. By 1970, a new phrase, 'pest management', had been borrowed from Australia and defined by B. P. Beirne in Canada as 'the reduction of pest problems by actions selected after the life systems of the pests are understood and the ecological as well as economic consequences of these actions have been predicted, as accurately as possible, to be in the best interests of mankind' (Rabb, 1970).

Pest management is therefore a blanket term for an approach to pest control which emphasizes economic and environmental considerations. It is really a definition of what pest control should and might be, rather than a practical protocol of the kind that the first 'integrated control' represented. Pest management equally embraces the multiple approach of integrated control and single component biological control in as much as either may prove the best solution to a particular pest problem. It might be a fair statement that integrated control is likely to prove the most generally applicable pest management solution. Indeed, the principal features of pest management listed by Rabb (1970) and summarized below have much in common with the features of integrated control already mentioned:

- The orientation is towards entire pest populations rather than to localized ones.
- The proximate objective is to lower the mean level of abundance of the pest so that fluctuations above the economic threshold are reduced or eliminated.
- The method or combination of methods is chosen to supplement natural control and give the maximum long-term reliability with the cheapest and least objectionable protection.

- The significance is that alleviation of the problem is general and long term with minimum harmful side effects.
- The philosophy is to 'manage' the pest population rather than to eliminate it.

The next stage in the development of the jargon was the definition in 1976 of 'Integrated Pest Management' (Apple and Smith, 1976). Apple and Smith accepted the term 'Pest Management' in relation to the control of separate groups of crop problems (insects, fungi, nematodes and weeds), but felt that the management of these different categories of problem should be integrated as a total crop protection system. The 'integration' of 'Integrated Pest Management' was therefore of disciplines rather than of methods, the latter being implicit in the term 'Integrated Control'. However, this definition of IPM was forgotten remarkably quickly; the terms 'Integrated Control', 'Pest Management' and 'Integrated Pest Management' are usually used as synonyms. One text on IPM (Burn, Coaker and Jepson, 1987) refers to it as 'a control strategy in which a variety of biological, chemical and cultural control measures are combined to give stable long-term pest control'. This definition is given as the simplest form, and the editors go on to the disclaimer 'we would like readers to form their own opinion as to what IPM is in practice'.

There has been one further development. In relation to management in agricultural and horticultural crops, there is increasing use of the phrase 'Integrated Crop Management' (ICM). The origins of this term are not clear, but the phrase emerged in Europe in the late 1990s. The IPM concept for crop protection solutions is a major component of ICM, but integrated with agronomy and ecosystem management. If farmers can be weaned away from routine spraying, then whatever progress research scientists can make in keeping pests below their thresholds by other control measures should be straightforward enough to integrate with an acceptance by farmers of the principle that insecticides should only be used when necessary.

Later developments of the integrated control concept: medical and veterinary pests

The term vector control has long dominated the literature dealing with medical and veterinary pests, but at a meeting on vector control in Khabarovsk, eastern Russia, in 1979 it was agreed that the future of vector control was with integrated control. In the 1980s the term 'Integrated Control' became increasingly used in respect of vectors, and in 1983 M. Laird and J. W. James

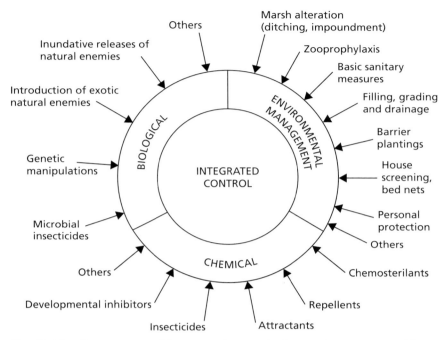

Fig. 13.11. Components of integrated control of mosquitoes proposed by Axtell (1979) and modified by the World Health Organization (1982) (courtesy of the World Health Organization).

published the first volume of two books entitled *Integrated Mosquito Control Methodologies*. They state 'For the present purposes we have considered "integrated mosquito control methodologies" as comprising chemical, biological and environmental procedures used conjointly or sequentially against a background of an exhaustive ecological understanding of the selected target pest or vector, so as to maximise efficacy, and be fully acceptable from health and environmental standpoints'. In the same year the WHO published a Technical Report Series called *Integrated Vector Control*. At first the emphasis was on biological and environmental control strategies, the latter usually becoming referred to as environmental management, which the WHO (1982) considered as a component of integrated control (Fig. 13.11).

In 1981, the WHO/FAO/UNEP and the United Nations Centre for Human Settlements established the Panel of Experts on Environmental Management for Vector Control (PEEM). The objectives of this panel are to create effective inter-agency and inter-sectoral collaboration between various organizations involved in public health, water and land development and the protection of the environment. Extended use of environmental management strategies for vector control within health programmes and in development projects could therefore be developed with minimum environmental damage. The terms 'Integrated Vector Control' or 'Integrated Pest Control' are still used by many. However, influenced by developments in the control of crop pests the term 'Integrated Pest Management' is now often used. Yet many, if not most, publications throughout the 1990s, including those from the WHO and other international organizations, persisted in referring to 'Vector Control' or 'Integrated Vector Control'.

It seems that medically important pests must be controlled and not managed. This belief goes back to a time when it was idealistically believed that diseases of humans could be eradicated (e.g. the actual eradication of smallpox, the failed eradication of malaria) and not just brought to a low incidence, i.e. 'managed'.

More recently the term 'Integrated Vector Management' (IVM) has sometimes been employed. This, however, seems an unnecessary complication, and the terms introduced at the start of this chapter – i.e. 'Pest Management' (PM) or even 'Integrated Pest Management' (IPM) – are suitable names for the integration of control measures whether aimed at agricultural, medical or veterinary pests and vectors.

Recently there has been renewed interest in IPM in medical and veterinary entomology. This interest has arisen for the same reasons that IPM developed earlier for agricultural problems, i.e. increased occurrence of pesticide resistance, more widespread use of pesticides and increased public environmental awareness. However, there is an important contrast here with agricultural pests. Whereas a variety of control methods can be considered for some medical and veterinary pests/vectors (e.g. mosquitoes), for some (e.g. flea vectors of plague) there are at present not the same range of possible alternatives to pesticides that can be proposed for crops.

Control versus eradication

The integrated control concept and its successor 'Integrated Pest Management' expects to control (or 'manage') pests and disease vectors to preconceived acceptable levels. Implicitly pests remain, albeit at ongoing low population levels, especially since biological control is usually one of the management components. There may therefore have to be some spraying or, if environmental or cultural measures are employed, then there have to be regular checks on their efficacy. As pointed out, cost–benefit calculations can be carried out in agriculture, where both inputs and outputs can be defined in hard cash, but in the medical field it may be difficult to decide what is an acceptable level of disease. For example, the stated aim of malaria *control* is to reduce transmission to an 'acceptable level', but what is this level? The usual answer given is that it is a level at which malaria no longer constitutes a major public health problem. To many this remains an evasive and unsatisfactory answer. Leaving this aside, maintaining control can be costly in terms of both money and resources, but in the control of agricultural pests the expense can be reflected in the selling price of the product.

In total contrast to the integrated control concept, eradication is the elimination of pests and vectors or of the infections they transmit. There is an absolute goal to be achieved – no more pests or disease. The American scientist Edward Knipling, the leader in the sterile male insect release method (see Chapter 9), coined the phrase 'Total Pest Management' (TPM) to rival IPM as another form of pest management.

The objective of TPM is usually far more difficult to achieve than to manage a population at a lower level, and far fewer people have heard of TPM than IPM! There have been, however, some notable successes, such as the eradication of the mosquito *Anopheles gambiae* from Brazil, of malaria from the USA, Taiwan, Seychelles, and other islands, eradication of dengue from Cuba, river blindness (onchocerciasis) from Kenya, the New World screwworm fly (*Cochliomyia hominivorax*) from the USA, several Latin American countries and from Libya (see Chapter 9).

Sometimes eradication has been followed by re-invasion of a pest or reintroduction of a disease. An example is the freeing of Mauritius from malaria in 1973, its re-introduction two years later, and its eradication again in 1990. In 1962 *Aedes aegypti* was eradicated from 22 countries in the Americas; today this mosquito has returned to all these countries except Canada, Chile, Uruguay and Bermuda.

In the long term it is usually cheaper, if it is possible, to eradicate a pest or disease than to control it ad infinitum. However, the situation has to be constantly monitored for detection of any re-introduction. For instance, the spread in many parts of the world of the Asian mosquito *Aedes albopictus* has been combated in some countries by rapid detection followed by eradication of this potential dengue vector, but the constant threat of re-introductions means there has to be continual vigilance.

Pest management packages

Although a synonymy between the words 'Integrated Control', 'Pest Management' and 'Integrated Pest Management' seems to have arisen in the minds of many people, the original concept of integrated control was largely directed at individual pest problems. Pest management in a crop, on the other hand, is usually geared towards a range of pests. The task of a pest manager is therefore to develop a package of control measures to deal with the pests of the crop as a complex, rather than each in isolation. Three approaches to pest management appear to have evolved.

Synthesis of target-specific controls (Menu systems)

The importance of pests in the cotton crop, its value and the fact that this crop is not destined for human consumption, has encouraged particularly heavy use of insecticides. It is therefore not surprising that the cotton crop has led the way both in the appearance of problems of pesticide use and in the development of pest management systems. The crop has justified enormous research inputs over a long period, with the result that a great deal is known about both insecticidal and non-insecticidal control measures for individual pests. Many researchers over a great number of years have worked on such measures, and much of the research stems from long before the scientists could have envisaged they were making a contribution to something that in the future would be called 'Pest Management'. An extensive arsenal of measures has thus become available for combination, in hindsight, to make packages appropriate for any particular area. This is not unlike making up your meal from the à la carte section of a restaurant menu. Diners need not all eat the same or even choose the same number of courses; hence the term 'menu systems' for such IPM systems.

The USA has had a long history of intensive research on cotton pests, and it is from the USA that the tradition of building pest management packages from individual components stems. The more important of these components for the cotton system are presented below in the order in which they have been discussed in the course of this book.

- Chemical control is very much based on operating crop scouting and economic thresholds. A chemical that was frequently used in response to threat from pests was endosulfan which, as mentioned earlier, does relatively little damage to many of the important parasitoids of the pests. Being an organochlorine, however, endosulfan is no longer used on cotton in the USA. Here GM cotton incorporating the *B.t.* toxin is increasingly used as an 'insecticidal' component of IPM.
- There have been several attempts to import natural enemies, and one of the main uses of biological control is the release of the parasitic wasp *Trichogramma*, an egg parasitoid of the bollworms *Heliothis* and *Helicoverpa*.

- There is a range of cultural controls, mostly enabling early harvesting and stalk destruction. Such cultural measures include uniform planting, early termination of irrigation and application of leaf defoliants and desiccants. As an additional cultural control measure, any adjacent lucerne is strip-harvested, as the presence of more mature lucerne strips means that any *Lygus* bugs are not driven from the crop to invade nearby cotton fields. Also, insecticide-treated trap crops may be sown as a control for boll weevil.
- Limited success against boll weevil (*Anthonomus grandis*) has been obtained with the sterile male release technique, as described in Chapter 9.
- Pheromones can be used in the 'confusion technique' to reduce populations of bollworm, and the aggregation pheromone for boll weevils provides an efficient 'lure and kill' trapping system for these beetles.
- A number of pest-resistant plant varieties is available. Early maturing varieties escape late-season bollworm attack; the 'Frego bract' characteristic gives some resistance to bollworm and high resistance to boll weevil. Varieties which are high in the secondary chemical gossypol as well as devoid of leaf nectaries provide further resistance to bollworm. Early maturing varieties allow early harvesting, which reduces bollworm problems.

The above type of pest management package has four particular characteristics. Firstly, it is extremely crop-specific; many of the measures relate to particular characteristics of cotton growing and could not easily be transferred to other crops. However, the plus side is that elements of the package may be transferable to cotton grown in another continent. Secondly, the control measures are not only crop-specific, but also highly target-specific. Each component is designed to reduce levels of one particular pest, and the components have been combined in such a way that the measures against pest A do not interact with attempts to control pest B. Thirdly, the major pest (bollworm) is attacked by a whole range of specific control measures; rather than 'killing two birds with one stone', a veritable basketful of stones is thrown at a single bird! Fourthly, as pointed out earlier, the package depends on a huge financial investment in many years of earlier research on single control measures against individual pests.

Another well-developed menu system is that for IPM in apple orchards in Europe and Canada. The need for such a system stems from the availability of organophosphate-resistant mite predators (*Typhlodromus pyri*) of another mite, the pest red spider mite (*Panonychus ulmi*). It then becomes important to preserve this biological control in spite of the necessity to control other pests, particularly Lepidoptera, with insecticides. The menu therefore includes selective materials for other pests that may be a problem. Such materials to control caterpillars include sprays of insect pathogens such as *Bacillus thuringiensis*, and insect growth regulators (see Chapter 3).

Computer design of menu-based packages

Much effort has gone into investigating the contribution that the modern ecological tools of life-table studies, systems analysis and mathematical modelling might make to the development of pest management systems. If the role of the various factors which cause changes in insect abundance can be understood and related to predictable events, then a model of the system enables predictions of the consequences of any pest control practices or combinations thereof. Ideally, the system could indeed then be 'managed' to best advantage, using computer technology.

The one obvious problem is that extensive life-table data must be collected over several years before a single pest population, let alone that of all the potentially important pests on the crop, can be modelled. Weather is an important determinant of population growth for all pests/vectors. For plant pests, there is the further complication that population growth is also affected by crop growth, nutrition and often cultivar. Elegant computer models of crop growth and the effect of environmental factors on crop growth and pest development have been devised for a number of crops, and computer models are thus already becoming useful for predicting pest incidence in the field. One of the most complete simulation models that have been developed to date is that for apple pests developed at Michigan State University in America. It modelled the production and utilization of assimilates in the whole tree, as well as the growth of leaves, shoots, roots and fruit. The tree model was driven by the environment via the input of weather data and was coupled with developmental models for some eight pest species under the acronym of PETE (Predictive Extension Timing Estimator system). One has the distinct feeling that the acronym must have preceded the association of words with the initials!

USA orchards make extensive use of natural enemies in the control of mite pests, and PETE incorporated four prey–predator models to help growers make decisions on maintaining biological control. When appropriate, monitoring assessments from the field were used to synchronize the model with the real world in order to improve the accuracy of the predictions of pest development. One of the great potentials that models possess is that they can include estimates of control costs and benefits and should allow decision making in relation to control alternatives on the basis of both short- and long-range consequences. It seems, however, that PETE never achieved widespread use, although it clearly attained a satisfactory level of prediction.

Such modelling is probably the most ecological approach to pest management, but the output of a model will obviously be only as accurate as the information with which it has been provided. Although models are already used to predict pest outbreaks in relation to economic thresholds (see above), a recent in-depth and worldwide review (Way and van Emden, 2000) bluntly concluded that models have as yet contributed little or nothing to control strategy in crops and even less to field practice. The important step towards the 'pest management' goal is not the development of the model in itself, but the early trial of its predictions in the field. Pest management 'output' can then be refined and improved in the light of experience in the 'real world'; up to now this step has been singularly lacking.

Protocols

In contrast to the menu approach, which usually involves selecting the components most appropriate for a given situation, IPM protocols require the package to be implemented in its entirety. Protocols are analogous to the 'set menu' in a restaurant. Protocols can therefore exploit synergistic interactions between control measures, and so the components are far less target-specific than in menu systems. Protocols are also rarely crop-specific, and thus their philosophy is probably more easily transferred to other crops in the same region than to the same crop in a different region.

Very simple protocols, following a common pattern, have been very successful in raising yields in developing countries (see the example of rice in Indonesia below). The simple and immediately practical philosophy is to introduce economic thresholds as the basis for spray decisions, and to recommend spray materials which are least disruptive to biological control (e.g. replacing pyrethroids with the older organophosphate compounds or endosulfan). Just these two steps often succeed in allowing biological control to make some contribution to the control previously obtained with insecticides, so that the intervals between the need to spray become progressively longer. At the same time, any obviously helpful cultural controls, such as provision of nectar plants for natural enemies or rapid destruction of crop residues, are introduced. Then, national research stations screen existing crop varieties to introduce some partial resistance against major pests. Such simple protocols can be introduced without the long time-scale and expense needed to develop menu systems, whether designed by man or by computer!

The principal mechanism underlying protocols is an indirect increase in biological control, i.e. they exploit interactions between biological control

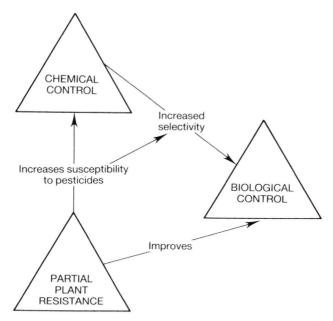

Fig. 13.12. The 'Pest Management Triad': the interaction of chemical control, biological control and host resistance.

and other control methods, as expressed in the 'Pest Management Triad' (Fig. 13.12). The underlying principle is that, if insecticide use is to be reduced, its effect must be replaced by another control component, the most generally available of which is likely to be biological control by indigenous natural enemies.

Partial plant resistance can often be shown to improve the impact of biological control; there are several reasons for this. Pests breed more slowly on partially resistant varieties, giving natural enemies a greater chance of exerting control. Moreover, as the pests are usually smaller, each natural enemy can consume more individuals before becoming satiated. A very important mortality from natural enemies additional to their feeding on pests is that their movements disturb pests and cause them to fall from plants. The proportion of insects falling is much higher on partially pest-resistant plants, on which the pests are more restless.

Partial plant resistance also stresses the pest so that it becomes less tolerant to insecticides. This means that the same kill of the pest can be achieved at a reduced dose, and this in turn improves the selectivity of the application in favour of natural pest enemies (see p. 299). However, the ability to reduce the dose of insecticide required by growing a partially resistant variety has a much larger effect than discussed earlier in relation to dose reductions. Natural enemies are not affected in the same way as the pest by plant resistance, and their insecticide susceptibility is likely to change relatively little when feeding on prey on a resistant variety. Thus, on a partially resistant variety, the dose–response curves for carnivores and herbivores separate out in such a way as to increase the selectivity window dramatically (Fig. 13.10b). The effect of plant resistance on the insecticide tolerance of the pest therefore promotes the impact of biological control still further. The final part of the triad is to utilize all the possibilities of achieving greater selectivity of a broad-spectrum insecticide discussed above (p. 295).

Using the philosophy of the Pest Management Triad, protocols can be developed quite rapidly by an experimental method, although the protocol may then be subject to considerable further refinement and improvement in later years. In contrast with the synthesis approach, and to some extent the computer design approach, the characteristics of the experimentally derived protocol are as follows. Firstly, it aims to tackle the whole pest complex from the start, allowing field results to determine the structure of the package rather than previous detailed knowledge about the agronomy of the crop, pest biology and population dynamics. Secondly, the package seeks to exploit the interactions that can occur between control methods, rather than keeping the inputs highly target-specific and free from interaction.

The practical implementation of such an experimental sequence is illustrated in Fig. 13.13. Several criteria can be used to evaluate each stage of the field trials. Since the entire pest complex is involved, harvestable yield is an obvious criterion; so is measuring the intervals between any sprays that become necessary. However, as there is often a trade-off between high efficiency of a single method (e.g. a pesticide) and biological control, other useful criteria include monitoring parasitoids and predators in the pest population as it recovers after the control intervention. In choosing the plant variety on which the package is to be based, one cannot expect resistance to the total pest spectrum. At this point a decision has to be made to target the plant resistance against either (a) one or more insects which account for the bulk of insecticide use or (b) the pests at a particular stage in crop phenology, e.g. seedling pests, so as to delay the first application of insecticide to the crop for as long as possible. After the selection of a variety, the remainder of the programme involves a sequence of pesticide experiments. Rather than relying on detailed foreknowledge of pest ecology, then, these experimental results can guide us to the best pest management process.

The rationale behind this sequence of experiments is that the number of experimental IPM combinations theoretically possible is enormous and

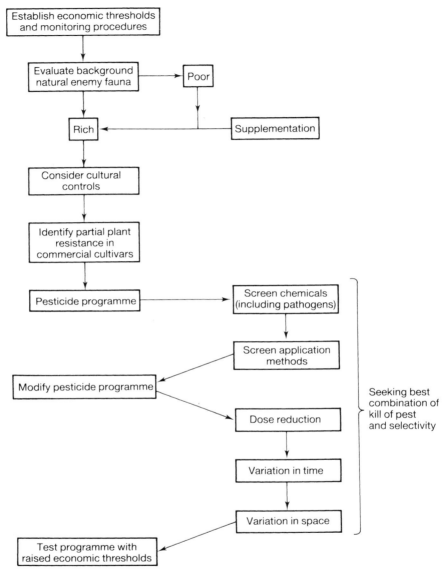

Fig. 13.13. Conceptual framework for developing a pest management protocol based on experimental variation of crop management.

beyond the scope of field experimentation. By taking the experimental programme in defined steps, each step comprises a limited number of treatments
for comparison. It is thus a necessary practical step that the order in which the
experiments are done and the decisions made after each experiment eliminate
a large number of possibilities which will never be evaluated. Conducting the
experiments in a different sequence would almost certainly produce a different
final pest management package; the aim is simply to reduce reliance on routine
pesticide use in a relatively short time by a programme which is applicable to
most crops in most situations.

The above account of pest management packages discusses approaches that
have evolved on crop systems. As stressed earlier, (a) no equivalent variety of
non-insecticidal menu or protocol components is available to medical and
veterinary entomologists, (b) usually a far higher level of pest reduction is
required than on crops and (c) the insects involved are mostly wide-ranging in
environments not as uniform or unitized as blocks of monoculture. However,
this does not mean that populations cannot be modelled and, in spite of
the doubts expressed above on what this approach has contributed to pest
management in crops, this contribution may greatly increase in the future. It
could well be the way in which pest management can be made more relevant
in medical and veterinary entomology. Progress to date in this area is described
below, though of course one can as yet only speculate on how such models
might be used in IPM.

Modelling medical and veterinary pest populations

Mathematical modelling of vector-borne diseases began in 1904 with Ronald
Ross who wrote a paper entitled 'The logical basis of the sanitary policy of
mosquito reduction'. Little attention was then paid to modelling until the
1950s, when George Macdonald applied mathematics and simple models to
the transmission of malaria and other diseases. Since then the malaria model
has been refined and models have been developed for a few other diseases and
vectors. The main aim of these models was to explain how diseases spread
amongst communities, that is disease epidemiology, but they also identified
the parameters most sensitive to control. For example, modelling has clearly
shown that the biggest reduction in malaria transmission occurs not when
vector populations are reduced in size, but when their adult daily survival
rate is shortened, sometimes by just 1–2 days. Today sophisticated stochastic
modelling is possible, because the power of computers allows variables which
are density-dependent and other population regulatory factors (see Chapter 2)

to be incorporated into a simulation model. Such models mimic changes in pest population size and disease incidence that would be caused by specified control measures. The aim is to tell managers when and where control should be implemented with a prediction of the likely outcome. A few successes have been claimed, including against house flies, some ticks of veterinary importance, mosquitoes and simuliid black flies. But as for crop pests, models have not yet provided much practical help in improving control programmes.

The trouble is that the baseline data for medical/veterinary insects is often of dubious quality. Also, just as for crop pests, relatively long-term field studies are often needed to obtain reliable estimates of various biological parameters, yet funding bodies want 'quick fixes'. There is additionally failure to recognize that the behaviour of the insects involved is even more heterogeneous than that of crop pests, yet in the computer all the individuals are usually (but see p. 290) programmed to behave in the same way.

Another problem, which applies generally to modelling in applied entomology, is that most modellers have little or no experience with the diversities and complexities of field situations. This has led some field workers to declare that mathematical modelling is nothing more than a game played by those who have no access to the facts. The counter argument is that even simple models may be able to unravel the complexities of species interactions and help focus attention on the most important biological parameters, for instance (as just mentioned) showing that reducing vector longevity is the key component for reducing malaria transmission. In summary, we feel that modelling will gradually become more realistic and will certainly have a role in our IPM armoury. This is provided that the models are tested in the real world in order to identify where our knowledge is incomplete or faulty as much as they are used to replace the labour of doing sums on the back of an envelope!

Conclusions

Chapter 5 sought to explain that the biggest problem of pesticide use is the development of pesticide-resistant pest populations, and that this is proceeding faster than the development of new approaches to chemical control. Most single alternative methods have problems which prevent them becoming general alternatives to pesticides, particularly in relation to the several pests that often attack one crop, or the different pest species that transmit infections to livestock or humans. So, can we control pests without insecticides? The answer must be 'no', at least in the foreseeable future. But the aim should be to minimize insecticide usage and its associated problems while maximizing

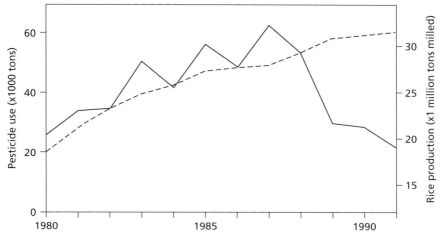

Fig. 13.14. Trends in pesticide use (solid line) and rice production (broken line) in Indonesia (1980–1991).

its benefits. Integrated control and its development to integrated pest management has sought to reduce reliance on pesticides by a multiple control measure approach; this has been most successful where the use of pesticides has augmented other control measures such as biological control and environmental control. Even so, farmers in intensive agriculture who can afford high insecticide inputs have not readily accepted such ideas as long as they still have effective insecticides available. Similarly it is regretfully much easier to try and tackle disease vectors with just insecticides than organize a programme comprising several integrated protocols.

The practice of pest management is, however, increasingly being promoted through a variety of economic and social pressures on practitioners. Some of the most important are:

- An association in overseas aid programmes for agriculture and vector control between provision of pesticides and the introduction of a pest management philosophy.
- Government pressure through subsidies and education in pest management. For example, although the Indonesian government 'decreed' in 1979 that pest management should be embraced, heavy subsidies for pesticides meant that overuse continued. However, in 1986 a Presidential decree banned nearly 60 broad-spectrum compounds. As a result, pesticide use between 1987 and 1991 fell by over 60%, accompanied by no reduction, and even an increase, in rice yields (Fig. 13.14). Since 1991, pesticide use

has fallen even further and by 2000 was probably only about 10% of the peak of usage in 1987.

- Pressure from supermarkets in relation to agricultural/horticultural produce. Supermarkets now account for about 80% of sales of fruit and vegetables in most developed countries, and are in a dominant position to insist which crop production and processing protocols are followed by farmers/growers. Much produce is now resourced from developing countries, and again the supermarkets seek to dictate the production conditions. Following pressure from the customers, the supermarkets not only seek to satisfy demand for so-called 'organic' produce, but also for produce grown under the best pest management technologies that are available for a given crop.
- The increasing role of consultants in taking pest-control decisions for crop and livestock producers, not only to advise them on crop protection, but also to take the necessary decisions and even to arrange for the operations involved. The staff of these consultancies are often trained in pest management, and seek to make their services attractive by reducing expenditure on crop protection by the maximum use of the pest management concept. There is of course always some pressure from single-agenda 'anti-insecticide' pressure groups. This is somewhat paradoxical because, by reducing reliance on insecticides, we should be able to preserve the current very useful and flexible arsenal of insecticides from the development of pest/vector resistance to them for far longer. Indeed, without availability of these insecticides, pest management might itself become an impossible goal.

The organization of the control of pests and vectors affecting humans is rather different from control of crop pests, and therefore changes result from different motives/pressures. The different organization of control arises because many control operations aimed at pests and vectors of infections of humans are directed by governments and regional authorities, or at the county or municipality level. Many of these organizations impose vertical control operations such as large-scale insecticidal spraying of houses to control malaria, Chagas disease and other diseases, or larviciding to prevent dengue outbreaks. There may be little political will to adopt an often more challenging integrated pest management programme. Moreover, large organizations often later receive reduced budgets or run out of funds, so effective control may not be economically sustainable in any long term.

With livestock pests the situation is again rather different. Although farmers often organize their own control measures, albeit following government guidelines, with very important diseases (such as tsetse fly-transmitted cattle trypanosomiasis and tick-borne swine fever), government or internationally organized control programmes may be enforced.

In the future there is likely to be greater use of technology in the control of agricultural, medical and veterinary pests and vectors. Molecular techniques are already used to improve toxicity of microbial organisms like *Bacillus thuringiensis* var. *israelensis*, and to incorporate genes encoding delta-endotoxin production into other organisms including plants (GM crops) that serve as food for pests. Commercial planting of GM crops on a small scale first began in 1996 (in the USA) and already in 2002 the world area sown to GM crops had risen to 60 million hectares. Similarly, genetic engineering is already being used to alter the genetic make-up of vectors of human diseases to make them prefer to feed on animals, or making them refractory to infection with disease organisms (as with vectors of plant diseases). However, there is a vast gap between the laboratory creation of 'harmless' vectors and getting these to replace natural vectors in the field.

Greater use will be made of the World Wide Web and computer technology, including satellite imagery, for pest and disease surveillance, to predict pest and disease outbreaks, and to coordinate control strategies. There needs, however, to be a sense of proportion and realization that any new approach is unlikely to be the panacea. Not only will new approaches bring new problems, as has been hinted in the relevant chapters of this book, but we also have every confidence in our insects to find their way round any new challenges they are presented with, just as they have done with the challenges of the past.

Appendix of names of some chemicals and microbials used as pesticides

The common name is given, followed by most frequent trade name(s) in parentheses.

Avermectins

Abamectin
Ivermectin (Mectizan)

Carbamates

Aldicarb (Temik)
Bendiocarb (Ficam)
Carbaryl (Sevin)
Carbofuran (Furadan, Curaterr)
Carbosulfan (Advantage)
Methomyl (Lannate)
Pirimicarb (Aphox, Pirimor)
Propoxur (Baygon, Aprocarb)

Chemosterilants

Amethopterine
Apholate

Fluoracil
Tepa
Thiotepa

Formamidines

Amitraz
Chlordimeform

Halogen compounds

Methyl bromide

Inorganics (including heavy metal salts)

Bordeaux mixture
Lead arsenate
Paris Green
Phosphine

Insect growth regulators (IGRs)

Cyromazine
Diflubenzuron (Dimilin)
Ecdysone
Fenoxycarb
Kinoprene
Lufenuron
Methoprene (Altosid)
Pyriproxyfen
Triflumuron

Microbials

Bacillus popilliae
Bacillus sphaericus

Bacillus thuringiensis
Bacillus thuringiensis var. *israelensis* (*B.t.i.*)
 (= *Bacillus thuringiensis* H-14)
Beauveria bassiana
Granulosis virus
Metarhizium anisopliae (Green muscle disease)
Nuclear polyhedrosis virus
Verticillium lecanii

Monomolecular films

Isostearyl alcohols
Lecithins

Natural organics

Azadirachtin (Neem)
Nicotine
Petroleum oils
Pyrethrum
Rotenone (Derris)
Ryania

Nicotinoids

Imidacloprid

Organochlorines

Aldrin
DDT
Dieldrin
Endosulfan
Gamma-HCH
HCH (BHC)
Isobenzan
Lindane (gamma-HCH)

Methoxychlor
Toxaphene (Camphechlor)

Organophosphates

Chlorphoxim
Chlorpyrifos (Dursban)
Coumaphos
Demeton (Systox)
Diazinon
Dichlorvos (DDVP, Vapona)
Dimethoate
Disulfoton
Fenitrothion (Sumithion, Fenitron, Dicofen)
Fenthion (Baytex)
Malathion
Methyl demeton (Metasystox)
Naled (Dibrom)
Para-oxon
Parathion, ethyl and methyl forms
Pirimiphos methyl (Actellic)
Schradan(e)
Temephos (Abate)

Organotin compounds

Triphenyltin

Pheromones

Codlemone
Muscalure (Flylure, Lurectron, Muscamone)

Pyrethroids

Allethrin
Alphacypermethrin

Bioallethrin
Cyfluthrin
Cypermethrin (Ambush)
Deltamethrin (Decis)
Esobiothrin
Lambdacyhalothrin (Icon)
Permethrin

Repellents

Bayrepel (Autan)
Citronella oil
DEET
Dibutyl phthalate
DIMP
Lemon eucalyptus oil

Synergist

Piperonyl butoxide

References

This bibliography includes references to items not cited in the text in addition to those that are. Those references in either category which we regard as useful general reading on pest and vector control are identified by the authors' names being highlighted in bold type.

Apple, J. L. and Smith, R. F. (eds.) (1976). *Integrated Pest Management*. Plenum, New York.

Axtell, R. C. (1979). Principles of integrated management (IPM) in relation to mosquito control. *Mosquito News*, **39**, 709–18.

Beales, P. and Gilles, H. M. (2002). Rationale and techniques of malaria control. In: *Essential Malariology* (eds. Warrell, D. A. and Gilles, H. M.), pp. 107–90. Edward Arnold, London.

Beaty, B. J. and Marquardt, W. C. (eds.) (1996). *The Biology of Disease Vectors*. University of Colorado, Colorado.

Birch, M. C. and Haynes, K. F. (1982). *Insect Pheromones*. Studies in Biology, no.147. Edward Arnold, London.

Burges, H. D. (ed.) (1981). *Microbial Control of Pests and Plant Diseases 1970–1980*. Academic Press, London.

Burn, A. J., Coaker, T. H. and Jepson, P. C. (eds.) (1987). *Integrated Pest Management*. Academic Press, London.

Carson, R. (1962). *Silent Spring*. Houghton Mifflin, New York.

Clark, G. C. (coordinator) (1994). Special Symposium on Vector Control. *American Journal of Tropical Medicine and Hygiene*, **50** (supplement), 1–159.

Copping, L. G. (ed.) (2001). *The Biopesticide Manual, 2nd edn*. British Crop Protection Council, Farnham.

DeBach, P. (ed.) (1964). *Biological Control of Insect Pests and Weeds*. Chapman and Hall, London.

Dent, D. (2000). *Insect Pest Management, 2nd edn.* CABI, Wallingford.

Doutt, R. L. (1958). Vice, virtue and the vedalia. *Bulletin of the Entomological Society of America*, **4**, 119–23.

Eldridge, B. F. and Edman, J. D. (eds.) (2001). *Medical Entomology. A Textbook on Public Health and Veterinary Problems Caused by Arthropods.* Kluwer, Dordrecht.

van Emden, H. F. (1969). Plant resistance to aphids induced by chemicals. *Journal of the Science of Food and Agriculture*, **20**, 385–7.

(1983). Pest management – routes and destinations. *Antenna*, **7**, 163–8.

(1987). Cultural methods: the plant. In: *Integrated Pest Management* (eds. Burn, A. J., Coaker, T. H. and Jepson, P. C.), pp. 27–68. Academic Press, London.

van Emden, H. F. and Peakall, D. B. (1996). *Beyond Silent Spring: Integrated Pest Management and Chemical Safety.* Chapman and Hall, London.

Flor, H. H. (1942). Inheritance of pathogenicity in *Melampsora lini. Phytopathology*, **2**, 653–69.

Gubler, D. J. and Kuno, G. (eds.) (1997). *Dengue Fever and Dengue Hemorrhagic Fever.* CABI, Wallingford.

Hadaway, A. B. and Barlow, F. (1951). Studies on aqueous suspensions of insecticides. *Bulletin of Entomological Research*, **47**, 603–22.

Hay, S. I., Randolph, S. E. and Rogers, D. J. (eds.) (2000). Remote sensing and geographical information systems in epidemiology. *Advances in Parasitology*, **47**, 1–357.

Hill, D. (1975). *Agricultural Insect Pests of the Tropics and their Control.* Cambridge University Press, Cambridge.

Knipling, E. F. (1955). Possibilities of insect control or eradication through the use of sexually sterile males. *Journal of Economic Entomology*, **48**, 459–62.

Laird, M. and James, J. W. (eds.) (1983). *Integrated Mosquito Control Methodologies, vol. 1.* Academic Press, London.

(1985). *Integrated Mosquito Control Methodologies, vol. 2.* Academic Press, London.

Lane, R. P. and Crosskey, R. W. (eds.) (1993). *Medical Insects and Arachnids.* Chapman and Hall, London.

Matthews, G. A. (2000). *Pesticide Application Methods, 3rd edn.* Blackwell Science, Oxford, and Longman, New York.

Maxwell, F. G. and Jennings, P. R. (eds.) (1980). *Breeding Plants Resistant to Insects.* John Wiley, New York.

Morgan, R. (ed.) (1992). *The BMA Guide to Pesticides, Chemicals and Health.* Edward Arnold, London.

Mullen, G. and Durden, L. (2002). *Medical and Veterinary Entomology.* Academic Press, Amsterdam.

Painter, R. H. (1951). *Insect Resistance in Crop Plants.* Macmillan, New York.

Panda, N., Khush, G. S. and Panda, K. (1997). *Host Plant Resistance to Insects.* CABI, Wallingford.

Pimentel, D. (ed.) (2002). *Encyclopedia of Pest Management.* Marcel Dekker, New York.

Rabb, R. L. (1970). Introduction to the conference. In: *Concepts of Pest Management* (eds. Rabb, R. L. and Guthrie, F. E.), pp. 1–15. North Carolina State University Press, Raleigh.

Ripper, W. E. (1955). Application methods for crop protection chemicals. *Annals of Applied Biology,* **42**, 288–334.

(1956). Effect of pesticides on balance of arthropod populations. *Annual Review of Entomology,* **1**, 403–38.

Rozendaal, J. A. (prepared by) (1997). *Vector Control. Methods for Use by Individuals and Communities.* World Health Organization, Geneva.

Samways, M. J. (1981). *Biological Control of Pests and Weeds.* Studies in Biology no. 132. Edward Arnold, London.

Schoeler, G. B. and Wikel, S. K. (2001). Review. Modulation of host immunity by haematophagous arthropods. *Annals of Tropical Medicine and Parasitology,* **95**, 755–71.

Service, M. W. (1993). Review. Community participation in vector-borne disease control. *Annals of Tropical Medicine and Parasitology,* **87**, 223–34.

(ed.) (2001). *Encyclopedia of Arthropod-transmitted Infections of Man and Domesticated Animals.* CABI, Wallingford.

(2002). (reprint). *Medical Entomology for Students, 2nd edn.* Cambridge University Press, Cambridge.

Southwood, T. R. E. (1973). The insect/plant relationship – an evolutionary perspective. In: *Insect/Plant Relationships* (ed. van Emden, H. F.), pp. 3–30. Blackwell, Oxford.

(1975). The dynamics of insect populations. In: *Insects, Science and Society* (ed. Pimentel, D.), pp. 151–199. Academic Press, New York.

Stern, V. M., Smith, R. F., van den Bosch, R. and Hagen, K. S. (1959). The integration of chemical and biological control of the spotted alfalfa aphid: the integrated control concept. *Hilgardia,* **28**, 81–101.

Tomlin, C. (ed.) (2000). *The Pesticide Manual, 12th edn.* British Crop Protection Council, Farnham.

Van Driesche, R. G. and Bellows, T. S. (1996). *Biological Control.* Chapman and Hall, New York.

Wall, R. and Shearer, D. (2001). *Veterinary Ectoparasites: Biology, Pathology and Control, 2nd edn.* Blackwell Science, Oxford.

Way, M. J. and van Emden, H. F. (2000). Integrated pest management in practice – pathways towards successful application. *Crop Protection,* **19**, 81–103.

World Health Organization (1982). *Manual on the Environmental Management for Mosquito Control: with Special Emphasis on Malaria Vectors.* Offset Publication no. 66. World Health Organization, Geneva.

(1983). Integrated vector control. *Seventh Report of the WHO Expert Committee on Vector Biology and Control. Technical Report Series,* **688**.

Index